*Twins*

*Also by Peter Watson*

War on the Mind

# TWINS

## AN UNCANNY RELATIONSHIP?

Peter Watson

THE VIKING PRESS     NEW YORK

Copyright © 1981 by Peter Watson

All rights reserved
Published in 1982 by The Viking Press
625 Madison Avenue, New York, N.Y. 10022

LIBRARY OF CONGRESS CATALOGING IN PUBLICATION DATA
Watson, Peter.
  Twins: an uncanny relationship?
   1. Twins—Psychology.  2. Nature and nurture.
3. Coincidence.  I. Title.
BF723.T9W37   155.2'34   81-51909
ISBN 0-670-73602-3   AACR2

Printed in the United States of America
Set in Times Roman

# Contents

# Acknowledgements

I should like to thank the many doctors, psychologists and sociologists who attended the 1980 conference of the International Society for the Study of Twins at Jerusalem, and were good enough to send me copies of their papers. This helped ensure the book was as up to date as possible; their names are mentioned at appropriate points in the text.

I have also made wide use of the following books: *Number Power*, by Keith Ellis (Heinemann, 1977); *Twins and Supertwins*, by Amram Scheinfeld (Pelican, 1973); *Lady Luck*, by Warren Weaver (Pelican, 1977). For details culled from magazine, newspaper and scientific journals, I am especially grateful to: 'Ethel Gaherty and Helen Moscardini', by Shirley Smith, *Littleton Independent* (Colorado), 9 October 1979; 'Twins reared apart: a living lab', by Edwin Chen, *New York Times Magazine*, 9 December 1979; 'The twins', by Cynthia Gorney, *Washington Post Style Magazine*, 10 December 1979; 'The twins', by Kathy Usborne, *States-Item* (New Orleans), 25 February 1980; 'Identical twins reared apart', by Constance Holden, *Science*, 21 March 1980; 'Reunion of identical twins, raised apart, reveals some astonishing similarities', by Donald Dale Jackson, *Smithsonian Magazine*, October 1980; 'Twins re-united', by Constance Holden, *Science 80*, November 1980. I am grateful to Peter Kellner of the *New Statesman*, and to Dr Hugh Gurling of the Institute of Psychiatry, for reading the manuscript, checking the statistics, and further suggestions. Chris Chippindale, my editor at Hutchinson, made many helpful comments.

Of course, without the help and cooperation of Professor Thomas Bouchard and his colleagues at Minneapolis, I doubt if the book could have been written at all. They are in no way responsible for the views and arguments expressed in the book, but to them, and to the twins themselves, go my warmest thanks for making *Twins* possible.

# Introduction
## *The 'spooky' coincidences between the Jim twins*

It all started with the 'Jim twins'. James Edward Lewis lives in Lima, Ohio, a town of 53,000 souls mainly employed in heavy industry: motor vehicles, steel castings, aircraft parts, machine tools. At the beginning of 1979, Jim Lewis was thirty-nine and working as a security guard. Twice married, he had had many jobs but by now was beginning to settle down. He was making good money and could afford a vacation in Florida every year. He did, however, have one unfulfilled ambition.

Jim had known since he was six that he was an identical twin. He had been separated from his brother when only a few weeks old after they had been adopted into different families. Jim desperately wanted to find his long-lost brother. His adoptive mother, Lucille Lewis, had been encouraging him and eventually, in January 1979, Jim started looking in earnest.

The Jims were born slightly prematurely on 19 August 1939 at the Memorial Hospital in Piqua, Ohio. Their mother, a thirty-five-year-old unmarried immigrant, immediately put them up for adoption. They were four weeks old when Ernest Springer, a utility company lineman, and his wife, Sarah, adopted one of the boys, whom they named Jim. They would have liked to adopt both children but, for some reason or other, they were told that their new son's brother had died during birth. In fact, Jess Lewis, a boilerman, and his wife, Lucille, adopted the second boy two weeks later. A court official recalls being flabbergasted when the Lewises gave their son exactly the same Christian name as the Springers had.

Jim Lewis's first move in the search for his brother was to

contact the court which had helped arrange their adoption. Before their adoption they had been looked after for a few weeks at the Knoop Children's Home in the small tobacco town of Troy in Ohio, but the home had closed, and there were no records to consult. The court proved particularly helpful, however, and after five or six weeks, Lewis's twin brother was found.

The reunion took place at Springer's home at Chelsea Avenue in Dayton, and both men were naturally a bit jittery. Lewis had contacted Springer by phone in advance – just to confirm they were indeed twins. Both men could scarcely credit that, after thirty-nine years, such a reunion could take place.

Lewis was late. He had stopped off for a last beer at a Pizza Hut to shore up his courage. Springer chain-smoked: did this – this . . . *twin* – need a kidney transplant? he wondered. Or money? There had to be some reason for this meeting. The knock on the door finally came.

Springer opened it, his wife Betty at his side. Lewis's fiancée, Sandy, was standing at the front of a line of people: Lewis's adoptive brother Larry, his sister-in-law Marie – finally, at the back, Lewis himself. In Springer's words, 'Jim and I shook hands and then we all just looked at each other for a few seconds. Then we broke out laughing.' Lewis said, 'Right off the bat I felt close, it wasn't like meeting a stranger.'

That was an understatement. For the coincidences in the lives of James Lewis and James Springer are astonishing. Their identical Christian names were chosen by their adoptive parents, and scarcely count. But what about the rest?

Both had married a girl called Linda, divorced her, *then* married a second time, to a woman called Betty
Lewis had named his first son James Alan, Springer had called his James Allan
Both had owned a dog as a boy, and named it Toy
Both men had worked part time as deputy-sheriff; both had been employed by McDonald's, the hamburger chain; both had been attendants in filling stations
Both spent their holidays at the same beach near

St Petersburg in Florida (a stretch of sand only 300 yards
long); both drove there and back in the same kind of car,
a Chevrolet
Both bite their fingernails – right down until there is
nothing left
Both drink Miller's Lite Beer
Both have white benches built around the trunk of a tree in
the garden
Both have basement workshops and work in wood, building
frames and furniture
Both chain-smoke Salems
Both put on 10 pounds when they were in their teens, for
no apparent reason, and both took it off again later
Both enjoy stock-car racing and dislike baseball
Both have had vasectomies
Both enjoy doing the household chores at weekends
Both scatter love notes around the house

The coincidences don't stop there. Both grew up with an
adopted brother called Larry; both had the same favourite
subject at school, maths, and both hated spelling, and were
not good at it; both have the same sleeping problems and use
the same slang words.

They even have identical headaches – 'tension' headaches
that begin in the late afternoon and turn into migraines. In
both cases the headaches started when they were eighteen,
and now they use similar words to describe the pain. Springer
says, 'It feels like someone is hitting you on the back of the
neck with a two-by-four.' Lewis, 'It's centred on the back of
the neck and it damn near knocks me out sometimes – I can
always tell when my brother gets a headache by the way he
acts – sort of sluggish. I know the feeling.'

Both have had two confirmed or suspected heart attacks.
Both have developed haemorrhoids.

The Jim twins were, as Lewis puts it, 'tickled' to meet up
again and find they had so much in common. 'I'll start to say
something,' says Lewis, 'and he will finish it.'

## The Minneapolis study

Psychologist Tom Bouchard was more than tickled. Scientists at the University of Minnesota have been studying twins for ten years. The walls of their laboratories are covered with the photographs of more than 600 people, each one a twin. But the Jims were ready-made material for the one aspect the Minnesota team have never had the chance to explore: twins who were separated very early on in life and remained separated until after their formative years were over.

The Jims' reunion was widely publicized. The uncanny similarities in their lives attracted a lot of attention; Associated Press syndicated their story, they appeared on the Johnny Carson show and were written up in *Science*, the official journal of the American Association for the Advancement of Science. Bouchard was one of those who heard and read about their meeting. Within forty-eight hours he got his colleagues together, raised a few thousand dollars from the university research fund and tracked down Springer and Lewis. Since then, and as a result of the publicity given to the Jims, Bouchard and his colleagues have seen another sixteen pairs of identical twins, all of whom were separated before they were three and not reunited until their late teens at the earliest. In all cases except one the separation took place before six months. Another eighteen pairs of separated identical twins are scheduled for future testing, including one Chinese pair (brothers from Hong Kong and Shanghai).

The 'spooky' coincidences, as Jim Lewis calls them, have continued to multiply with each set of twins that has flown into Minneapolis. Two British women, Terry Connolly and Margaret Richardson, found that, although they were separated in 1943 and did not even know they were twins until 1979, they were married on exactly the same day in 1960, within an hour of each other. Two other British twins, Dorothy Lowe and Bridget Harrison, chose to fill out diaries for just one year in their lives – 1962 in each case. Not only did they buy the same make of diary, the same model and the same colour, they even filled out the same days of the year. Both played the piano as a girl, but gave it up at the same time; both like wearing flamboyant jewellery. A third pair, separated for over thirty years, found they had both fallen downstairs in the same year and injured their ankles. Can these coincidences be explained? Is there a special 'bond', a psychological or 'psychic' link, between identical twins? As a psychologist, Bouchard – together with most of his colleagues – is more interested in the differences between people. But he does admit now that he has been surprised by the similarities in the twins: he never expected to see so many. 'The scores on so many of the tests are incredibly close: closer than those of the same people taking the test twice.' Moreover, he doesn't think this is a bias he or his colleagues have acquired and which influences them to look for similarities. To avoid such bias they contract out some of the testing to professional testers who are not told what the aim of the experiment is. 'The similarities are there all right,' says Bouchard. 'You can see it, very often you can measure it, and when you meet the twins you can *feel* it in all sorts of ways you can't put into words.'

Or into academic papers. When you visit Minneapolis you hear all sorts of stories about things happening while the twins are there that might be coincidences – or might be something else. Like the way Bouchard deliberately books them into rooms in different parts of the hotel – to see how many times they bump into each other in the elevator because they have had a common whim to do something at the same time or are seeking each other out at the same time. Or the number of times the twins have identical reactions to things during their stay in the city.

## Coincidence, the 'twin bond' and politics

This book investigates these coincidences and the light they throw on the 'twin bond'. A fundamental problem is that there is no scientific definition of a coincidence and no systematic way even of measuring one. For instance, if two men – twins – turn up at Minneapolis airport wearing the same shirt, the same colour and the same design, then, if they have not met for twenty years or so, that seems spooky. If each Jim marries a Linda, divorces her and then marries a Betty, that seems spooky too. On the other hand, if women twins turn up at Minneapolis airport both wearing skirts, is that a coincidence? Or if two male twins both turn up without ties, is that a coincidence as well? It certainly is not spooky; does it count at all?

Further, we have no real way of knowing, in many cases, how unusual these habits are. We do not know how many men marry Lindas or wear blue shirts with epaulettes. Nor do we know how many coincidences there are in the lives of *any* two people who bump into each other in the street. Once you start looking, maybe there are more than you think. But no one can be certain. The way you count coincidences can be all-important, too. Jim Springer is still married to his Betty, but Lewis is now engaged to Sandy: how much does that weaken the coincidence in their marital lives?

What has actually been written about coincidences in the scientific field has been of two types. In the first place 'fringe' scientists have written rather idiosyncratic accounts. Carl Gustav Jung wrote *Synchronicity*, a study of simultaneous events and the rhythms that run through certain series of happenings. This book is rather old-fashioned in tone, a mixture of mysticism, astrology and philosophy. Paul Kammerer, an Austrian biologist of the early twentieth century who is chiefly remembered for his fraud over the midwife toad, discovered a 'law of series' in which coincidental events run in (say) threes. For example, a bus ticket and a theatre ticket (bought the same day) both bear the number 9, and they are soon followed by a telephone call in which the same number is again mentioned. Kammerer would spend hours, wherever he went, recording the height, hair colour and type of hat worn by every passer-by. He made the observation that men with red hair tended to pass by in clusters with long gaps in between. Another man with similar interests is Arthur Koestler, who has spent no little time collecting examples of striking coincidences and has drawn tentative conclusions about the way they may be organized. Several of these are discussed later.

The second type of writing about coincidences comes from mathematicians. They have not sought to *explain* coincidences but instead have worked out the odds against some of the more exotic ones occurring by chance; this helps one to judge whether paranormal phenomena could, at least in some cases, be responsible. In one instance, the fifteen members of a choir were *all* late for practice one night – fortunately as it turned out, since the chapel blew up a couple of minutes after they should have started. An act of God? Mathematicians, armed with the statistics of punctuality, can work out what the odds are for fifteen people being late on the same night (see page 103 for the answer). Attaching probabilities to events, as bookmakers do, gives a relative bench mark which helps us understand exactly how infrequent an apparently rare event is. (We can compare it, for example, to being struck by lightning or getting a full house at poker.)

In many other universities and hospitals twins, and the 'twin bond' particularly, are coming under scrutiny. The International Society for the Study of Twins, formed as recently as 1974, held its third conference in June 1980, in Jerusalem. Research papers were given by academics and doctors from France, Germany, Norway, Finland, Israel, Great Britain, Singapore, Australia, Japan, the USA and Nigeria. In most of these papers some aspect of the bond between twins was being examined. And all sorts of coincidences between twins were reported.

There are three reasons for this renewed scientific interest in twins and the twin bond.

*With more enlightened attitudes towards illegitimacy, twins (and other children) are 'adopted out' less these days than in the past; it follows that now may be our last chance to search out twins who have been separated and raised in different circumstances. This natural social laboratory may not repeat itself.

*The increasing use of computers in medicine has meant that there are many more comprehensive registers – of twins, of cancer patients, of death records and so on – which enable large-scale comparisons to be made.

*Probably most important, the growth of the behavioural sciences in the 1950s and 1960s carried with it the firm conviction that the environment is mainly responsible for shaping psychological and medical development. Intelligence, personality, mental health and illness, not to mention physical health and illness, were seen as owing most to the quality of mothers, the quality of schools, the quality of housing and so on. Lately, however, doubts have grown. These factors remain important (they are in any case the only things we can try to change), but many scientists now suspect that genes play a much bigger part in our lives. And genes may affect our psychological as much as our physical make-up.

Being at the centre of the 'nature–nurture' debate has meant that twin research is no stranger to controversy. As long ago as 1936 the Russians banned twin studies as potentially in conflict with Marxist theory, and they were not 'rehabilitated' until the early 1960s. More recently the twin studies of Sir Cyril Burt, which purported to show that intelligence was largely inherited, have been shown to be fraudulent, with invented calculations and even fictitious 'collaborators'. Since the 1960s any fresh result which has shown or purported to show that our genes exert some influence over our psychological, as opposed to our physical, make-up, has been seen as sinister, and the scientists who reach these conclusions said to have uncomfortable (and invariably right-wing, Fascist, or other ultra-reactionary) political views. This is not at all how many geneticists, and others involved in twin research, seem to me. Few have any overall view of 'human nature' stemming from their research. Even where they do think that genes are very important and people consequently less malleable under environmental influence, it does not follow they are more 'right-wing'. Just as often their knowledge leads them to a more accurate idea of the way the environment could be beneficially changed. For instance, Dr Richard Rose at the University of Indiana has been studying the children of identical twins (see page 180). Legally these children are cousins, but biologically they resemble half-brothers and half-sisters since one each of their parents have identical genes. Now Rose has found that the children of female twins are closer, psychologically, than the children of male twins. In other words, there appears to be something about 'mothering' as opposed to 'fathering' or 'family influence'. Since most studies investigating the effect of class define families according to the father's occupation, and the mother is neglected, who knows what influences have been overlooked? This is a perfect example of a *genetic* study throwing light on the workings of the *environment*. Yet many psychologists and sociologists seem unwilling to acknowledge this. And let no one doubt that they have friends in high (scientific) places. The leader of the team of scientists at Minneapolis, Tom Bouchard, believes that they have been denied funds at least once because of the potentially 'political' nature of their results. It is one reason why they have been slow to publish their findings in the scientific journals: they want to collect a lot of evidence first, rather than risk controversy which may disrupt their study.

However, I would bet that most ordinary people are intrigued by the coincidences at Minneapolis, and not just for trivial and superficial reasons. They sense that underneath lies one very important point.

If the coincidences really are so unusual, they might be due to paranormal phenomena, extrasensory perception, astrology or something similar; in which case the scientists' results must cause us to rethink our whole conception of human nature. Or we are dealing with an extraordinary revelation about heredity – our genes are controlling even minute aspects of our behaviour over as long as thirty or forty years. This is not quite as sensational as the rethink required if ESP actually exists, but not far short.

Alternatively, the Minnesota study shows no such thing, and the coincidences are nowhere near as surprising as they seem, either because the behaviour is not as rare as it first appears or because the scientists have been searching too hard for coincidences and have exaggerated the significance of the number found in the lives of twins.

Whichever of these it is, the answer can scarcely fail to be fascinating: it is curious to think that so much depends on the names of two wives, on the entries in two cheap diaries, or two marriages celebrated the same day. But it is true: everything depends on the coincidences.

## Types of twins

Two other points need to be made before we can proceed to a detailed description of the lives of the twins who have been studied at Minnesota. First, we should be clear about the different types of twins.

A twin pregnancy may happen in one of two ways. In human beings, one egg is normally released at each ovulation; if fertilized, it grows into a single foetus. If two eggs happen to be released, both may be fertilized, independently and by different spermatozoa. They grow into two separate foetuses, each with its own placenta, which are born as 'fraternal' or 'non-identical' twins; the scientific name for them is dizygotic, from the Greek for 'two eggs'. Dizygotic (DZ) twins happen to be conceived and delivered together, but they are genetically no closer and look no more alike than any brothers or sisters; often they are of different sexes.

Identical twins start from a single fertilized egg which, instead of developing into a single foetus, divides into two separate individuals. The scientific name, monozygotic or MZ (from the Greek for 'one egg'), emphasizes that identical twins have identical sets of genes. They are always of the same sex and similar in appearance (although we shall see that some pairs of 'identical' twins are more identical than others).

Occasionally identical twins do not separate entirely in the womb, and are born with parts of their bodies fused together. These are called 'Siamese twins' after the famous brothers Chang and Eng.

It follows that all differences between MZ twins must be due to the effects of their environment, whereas the differences between DZs can be due either to environment *or* to their genes. If more MZ twins than DZs share a disease (like heart attacks or schizophrenia), then to that extent these illnesses are genetically based. This is why the Minnesota study is important. If the coincidences are not an artefact of the investigation and they are not due to extrasensory perception (ESP) or other 'psychic' powers, then they suggest that our genes affect our lives in all sorts of ways – large and small – hitherto unsuspected.

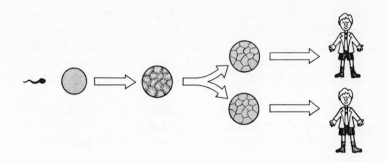

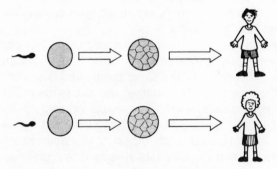

Monozygotic (MZ) and dizygotic (DZ) twins

## Four reasons for caution

I think the Minneapolis experiment is interesting and important. So do many other scientists. But we should bear in mind that there are many criticisms one can make of twin studies, quite apart from any difficulty there might be over defining and measuring the coincidence.

Susan Farber, a psychologist at New York University, has done science a great service by re-analysing all the *previous* studies of identical twins reared apart (her book* came out in the United States as this one went to press but has not yet appeared in Great Britain). Before the Minneapolis study there had been several others which looked at the same

*Identical Twins Reared Apart* by Susan L. Farber, New York, Basic Books, 1981.

subject, though some had reported on only one or two pairs of twins. In all, Dr Farber found 121 sets of MZ twins raised apart, and her conclusions about these studies make sobering reading. We shall come back to her arguments at several points in what follows but, briefly, her main strictures are these:

*Many twins in raised-apart studies were found by public appeal, over TV or newspapers, and were therefore confined to twins, of whom one at least knew that he or she was a twin.

*In many cases the twins had found out they were twins because they had been mistaken for the other. So *by definition* many of the twins in twin studies are those who are similar to each other, at least physically.

*On examination, it turned out that many twins (90 per cent) had not been truly separated. Very often they had been raised in different branches of the same family and occasionally had seen their twin. In fact, out of the 121 twins in Dr Farber's combined sample, only *three* could be said to be truly separated in the sense that they had been split up as babies, hadn't known about each other in the intervening years and had been seen by scientists at their first reunion.

*By definition, these twins were the product of broken homes. This, combined with the fact that most of them were split up either because their families were too poor to keep them, or because they were born illegitimately in war-time, or both, makes them not at all typical of people in general or even, come to that, of twins.

The Minnesota study answers some of Dr Farber's criticisms of earlier studies, but not all. The reader needs to allow for that throughout this book.


What follows falls into three parts. The first part describes in detail the Minneapolis studies, and gives a full account of the 'uncanny' coincidences found between the separated twins, in their dress, their behaviour, their medical problems, psychiatric symptoms and sexual practices. The childhood and home background of the twins are described so the twins can appear not just as 'subjects' in an experiment but as real people with families, feelings and attitudes. The second part examines the 'natural bond' between twins, those factors that make twins, in all sorts of ways, different from non-twins. It includes anthropological and mythological evidence about twins in history (which tends to show that, in most societies and at most times, twins have been an unwelcome addition to the family and tribe) and the medical and sociological facts of twin-hood. Identical twins not only share a common genetic make-up; they look alike, and are generally treated alike by their parents, dressed the same, given matching names, and so on. Before we imagine anything 'uncanny' about the bond between twins, we first have to allow for the similarity this natural bond leads to.

The third part is the heart of the book, analysing the 'uncanny bond' and trying to examine in an objective way the extraordinary coincidences seen at Minneapolis. The world is a large place, contrary to what we are often told, so rare events, unlikely happenings, extraordinary coincidences are taking place all the time; their occurrence is no reason to suppose unnatural forces are at work. In this section are reviewed a number of intriguing coincidences and some generalizations are found to be possible about them. I then go on to the *specific* aspects of behaviour of the Minneapolis twins, and attempt to assess just how unusual they really are. If we know, for instance, what proportion of people bite their nails, or fall downstairs in any one year, or play the piano, we can establish how unusual the coincidences observed between twins at Minneapolis really are. The next chapter takes account of behaviour genetics, now one of the most important areas of research in psychology, and attempts to answer the crucial question: compared with two strangers taken at random from the street, how much closer, *psychologically* speaking, are twins? Finally I introduce a fairly new science, chronogenetics, the theme of which is that genes affect not only how we are at birth but also govern our development, its spurts and lags, mentally and physically, *throughout* our life, up to and maybe including our death. Chronogenetics may explain why some twins not only share certain characteristics, but develop them at the *same time*.

And it is those similarities between twins, the ones which

occur simultaneously, that are the most interesting and the most puzzling, for me at least. It was the mystery behind these particular coincidences that made me want to write the book in the first place.

*Part One*
# Collecting Coincidences

# 1   *In the Twin Cities*

## *A surprise at the airport*

Jake Hellbach arrived in Minneapolis on a cold Sunday in March 1980. He had flown in from New Orleans, hours after his twin, Keith Heitzman, had touched down from Dallas/ Fort Worth. They were the thirteenth set of twins to arrive in the city for a week's intensive testing. Jake was met by Tom Bouchard and shown to the scientist's car, a 1978 Chevrolet. Before they had driven out of the airport Jake remarked on a slight shuddering sound on the nearside of the vehicle, towards the rear. It was a small knocking sound, nothing serious, and due to an uneven tyre. But that was not what made Bouchard smile. It was just that, earlier in the day, when he had picked up Jake's twin at the airport, Keith too had remarked on the *same* sound in Tom's car.

The twin cities of Minneapolis and St Paul are pleasant conurbations; they now comprise a single metropolitan area divided by a gorge through which flow the waters of the Mississippi. The grey-brown river is already swift even though it still has 2300 miles to go before it reaches the Mexican gulf. In spring and summer the towns are warm and sunny, with many leafy streets. But for five months of the year the area is covered with a thick blanket of snow and temperatures tumble. So much so that the University of Minnesota, on the southern bank of the river, is built half-underground. Long underground corridors, like streets from *1984*, connect the academic departments with the faculty club, the parking lot, the snowed-up football stadium.

When twins arrive to take part in the Minnesota study they are usually met at the airport by Tom Bouchard or one of his colleagues. The coincidence between Jake's arrival and that of his brother Keith was only one of several which have occurred during the study. A British pair of twins turned up wearing identical jewellery, and twin men, separated for twenty-five years, arrived at the airport wearing identical shirts – the same colour and the same style – and identical spectacles.

Yet . . . though he has been intrigued by these similarities at the airport, and though they have undoubtedly had an effect on him and his feelings about twins, Tom Bouchard is not keen to attach too much importance to what happens in the baggage claim area.

Twins who take part in the programme generally arrive in Minneapolis on a Sunday, unless they have come from Britain (the four pairs of British twins who have so far visited the research study have arrived on Saturdays to give them time to get over jet lag). On Sunday night Bouchard has dinner with the twins and explains the study to them. He tells them what to expect during the coming week, and warns them that they will be exhausted by their time in the twin cities. They sign consent forms and the study proper is ready to begin. Bouchard has taken a few notes by now, observing whether the twins have a tendency to sit in the same way or use the same gestures, whether one is dominant, the other submissive. Nothing conclusive, just preliminary observations before the twins have the chance to influence one another.

## Questions, questions, questions

The study proper generally begins on the Monday morning and runs through, almost nonstop, until mid afternoon the next Saturday. Evenings and the whole of Wednesday are free but the other working days run from 8 a.m. until 5.30 p.m., a tiring schedule. During that time the twins answer roughly 15,000 questions about themselves and see several psychologists.

### Medical history and medical evaluation

First, the twins give an extensive medical history. They are encouraged to think of all the childhood illnesses they have had, when they had them, how sick they were and so on. 'We want to know whether they've had mumps, broken legs – everything. When, how long and how serious!' says Bouchard. They have a physical examination, and they are also asked to bring any medical records they have with them. Even birth certificates or school records may be valuable for the extra information they contain. A test of heart and regulatory systems follows, including a cardiogram. This is carried out at the University's Varsity Club Heart Hospital and for a day the twins wear a heart monitor around their necks. Since many of the twins are approaching middle age, certain abnormalities which are developing are sometimes spotted in these tests. For example, Barbara Herbert and Daphne Goodship (British twins) both discovered in Minneapolis that they were developing identical heart murmurs, although the murmurs turned out to be nothing serious. A ninety-minute allergy test involves blood samples and some skin tests. Next comes an examination of the lungs, especially their capacity – this can be very useful in research, especially in regard to smoking. (I report on some interesting results regarding smoking in chapter 6.) On one day the twins take a specified breakfast, which enables the doctors to test whether or not they are diabetic. Finally on the medical side, two doctors take details of the twins' sex lives, the aim being to explore to what extent sexual behaviour and sexual responses are genetically based or whether they are learned. This addition to the study is relatively recent, and few results as yet are available.

*Physiological assessment*

Professor David Lykken, one of Professor Bouchard's colleagues, tests the twins' heart pattern, the resistance in their skin, and their brainwaves. In order to do this their scalps have to be rubbed clean so that electrodes can be fitted. This particular manoeuvre made Jake Hellbach nervous – but Lykken reassured him: 'I'll give you a record to take home: just to prove something's going on in there.' This research has already provided a number of intriguing sidelights. For instance, in order to have their brainwaves tested, the twins have to enter a small sound-proofed booth, rather like a telephone kiosk but without the windows. Dr Lykken has found that a number of MZ twins show the same reaction to the booth itself, irrespective of the test they do in it. If one twin feels claustrophobic and is wary of entering the booth, the other twin usually has the same reaction. And where one twin does not mind the booth, the other twin has no objection either. Lykken has found also that the brainwave patterns of twins are *more* similar when they are apart than when they are together in the same room. This is an interesting result, the more so as the scientists were neither looking for it nor expecting it. Lykken also takes the twins' fingerprints, handprints and footprints (he tends to find that twins grow callouses on the same parts of their hands and feet). He measures head size and weight, and takes photos of all the twins.

*Psychological and psychomotor tests*

The twins take intelligence tests, tests of verbal fluency, perceptual speed, memory, and a variety of personality tests. Questions like: 'How many uses can you think of for a brick?' are asked. Twins have to draw a house, a tree and a person (to reveal different aspects of their personality). Their common dreams are noted, how well they take decisions and so on.

Next there are various psychomotor tests, for example, a hand-steadiness test and a hand–eye coordination test (men seem to perform better than women). The twins also spend some time on a treadmill (literally) to see how they react to sustained activity. This prompted Jake to say he felt like 'the $6 million dollar man'. Bouchard spends the Friday morning taking the twins through their life with special emphasis on how they have reacted to stresses and crises – exams, for example, or bereavement, divorce, failing to get promotion, or being fired.

Another probing interview explores the twins' personal lives – their feelings, attitudes, significant experiences and so on. This is supplemented by an interview with two of the investigators which is videotaped. This is not especially personal – though the twins are asked to bring along diaries, schoolbooks, anything that will help jog their memories. The videotape has two purposes: to enable the scientists later on to look at the gestures and mannerisms of the twins and search for similarities; and also to enable other scientists to look at the twins and make their own independent judgement about them. Finally, the twins' handwriting is sampled, and their spouses, if they are accompanying them, undergo some of the psychological testing.

As with the brainwave tests, Bouchard and his colleagues have found that twins tested in different rooms produce results that are *more* similar than when they are together, albeit 'back-to-back', in the same room. Secondly, many twins not only get the same scores on the tests but they make the same mistakes and finish at much the same time. This sort of detail has made a deep impression on some of the twins who have, quite definitely, found out at Minnesota that they were more alike than they thought. They have discovered interests, preferences, attitudes in each other that, by being mirrored in someone else, make them understand themselves better. Making the same mistakes on the tests, as much as getting the same overall score, has had a big impact on several pairs of twins. It is as if seeing another's shortcomings has finally convinced them of their own. This kind of reaction is difficult to analyse scientifically, but it has excited Bouchard and his colleagues.

# 2  Lives of the Twins

## Oskar and Jack: Nazi and Jew

Perhaps no story in this book is as uncanny and as perplexing as that of Oskar Stohr and Jack Yufe, identical twins born in Trinidad in 1933. Oskar was raised as a Nazi, a member of the Hitler Youth; and Jack was brought up a Jew, and worked in two kibbutzim in Israel. Yet when they went to Minnesota to take part in the twins study the coincidences between them were truly astonishing. But since the story is so strange, let us go right back to the beginning.

In 1933 Trinidad, which lies at the southern tip of the Caribbean, was a British colony. It had not yet discovered the offshore oil for which it is chiefly known today. The economic situation on the island was not good and for this reason there was trouble between the twins' mother and father almost as soon as the twins were born, so much so that, by the time they were six months old, they had been separated. Their mother, who was a German and a gentile, took the first-born twin, Oskar, and an older daughter back to Europe. Jack stayed in Trinidad with the father, a Jew. As they grew the brothers never met or wrote to each other, although they were aware that they were twins.

Oskar was raised by women, about fifteen miles from the German–Czech border, in a small town of about 8000; there were mountains and forests near by. Shortly after she returned to Germany, Oskar's mother left for Italy to work and the boy was raised by his grandmother, who was helped, later, by his older sister. It was not a particularly easy life but it was fun. Grandmother Stohr was a small, blonde woman, soft-voiced, a devout Catholic who grew her own potatoes, rhubarb, strawberries and blackcurrants in a vegetable plot near the house. Oskar had to help with this vegetable patch and also do other chores such as chopping logs. Some aspects of his existence were very hard. For example, every Sunday he was sent, without fail, to church. In that part of Europe it can get very cold in winter and, in the late 1930s, the church was not heated. But he always had to go and, when he was nine, he received his first communion: the occasion was celebrated with a new suit, made of special dark-blue cloth his mother had sent from Italy.

He was ten when the events of the outside world first encroached on the young boy. He remembers seeing a newsreel as a child: it was about Jews in the Polish ghetto. 'They were thin, gaunt, with beards. With their black *yarmulka* caps, they were frightening to children.' Suddenly, cut into the newsreel were pictures of rats scurrying around. 'They were compared to vermin; we were told they multiplied like rats and, like rats, had to be exterminated.' Then one day the headmaster of his school asked to see him. The head watched carefully as Oskar entered his room and gave the customary salute: '*Heil Hitler*' (this was in 1943). The headmaster reproved Oskar, saying that his '*Heil*' was shoddy, his arm not extended crisply enough. 'And you should snap your heels together as you shout the greeting,' he snarled at the boy. 'Go out and come in again.'

Oskar did as he was told but the headmaster still seemed suspicious. He went on to ask about Oskar's older sister who, because their parents had been together longer in her case, had kept her father's name, 'Yufe'. The head said that it sounded as though it meant '*Jude*' – Jew.

Oskar recalls today that he had to do a bit of quick thinking. 'It's a French name, sir,' he said. 'There's an accent over the "e".' In truth he had no idea where the name came from. But still the headmaster kept on. Where was Oskar's father? In South America, said the boy, but he did not know exactly where. If he had said his father was in Trinidad it might have been taken as meaning he was a British subject – almost as bad as being Jewish.

The headmaster seemed appeased, however, for he then asked the child whether he was related to another Stohr in the area: this man was a Nazi stormtrooper and a well-known local athlete. Oskar was relieved to admit that this man was his uncle. The interview was over.

A surprise was in store for him that night, however, when he got home. He told his grandmother what had happened, and she decided the time had come to tell him a family secret she had kept to herself all these years: his father in Trinidad had indeed been a Jew.

It was clear to Oskar, even at that age, that this was a secret he would have to keep for a very long time. Two Jewish families had already been taken by stormtroopers from the street in which he lived. In one case the wife and children had been taken during the day and when the father came home from work at night the stormtroopers were waiting for him; they had beaten him with their rifle butts and then taken him off. The lesson was not lost on the ten year old.

In fact, he became a good student at school and a popular one. He was good at sports, especially swimming and athletics, but it was political education that was central to schools in Germany at that time and even there Oskar excelled – that is what makes him so interesting. Years later he was to remember that schoolchildren had to learn various sets of instructions about Jews. Oskar was very good at learning things by heart and, at one point, was actually commended for having remembered the instructions against Jews so well.

When first interviewed in 1979 about this, Oskar found it difficult to explain but said that he had always thought of himself as a Catholic and so did not take a lot of the lessons about Jews personally. He learned about the Treaty of Versailles, that it was dishonourable, that the Jews had been responsible for it and that Hitler had reversed it. But, he says, he understood it to have been about religious differences, not about racial ones.

It became a question of race, of course, as even Oskar was forced to acknowledge. But, when asked whether it was fear that prevented him linking anti-Semitism with himself or his father, he replied, slowly, 'That could have been the reason.' And, he added, 'Maybe the reason I learned the [anti-Jewish] lessons so well was because I had so much knowledge.'

Only after the war, says Stohr, did the full horror of what

had gone on come home to him. The part of Germany in which he lived was occupied by the Russians, and food rationing began. One day all the Germans in his village had to line up in the school gymnasium. Then they were told – and shown – about extermination. 'There was an exhibit in the gym – boxes – of bones – one box with human hair – and boxes with clothing – and – er – lots and lots of pictures. There was an exhibition of pictures from the concentration camps and everybody was made to go . . . everybody had to walk through the corridors . . . all the Germans were forced to go there.' The exhibition 'portrayed it as fate, and that it was not only the fate of the Jews but also the fate of the Czechs . . . not just the Czechs themselves but any enemy of the Nazis . . . some women who couldn't bear to look had screaming fits. It was awful.'

Did he think to himself, as he saw those boxes, 'My people did this'?

Stohr nodded.

How did that feel?

'I couldn't really grasp the thought. On the one hand it could have been my fate; but on the other, it was also my people who did that.'

It is a moving story. At this distance, it is perhaps not necessary to judge. Many other Jews found themselves in similar situations. Some spoke out and identified themselves and were shot. One whole family, as we now know, lived the entire war in Frankfurt by posing as a gentile family (and the entire street knew). One man who does not condemn, does not judge Oskar, is his twin brother, Jack Yufe.

The *Copper Fin* played a large part in Jack Yufe's early life. It was his beloved rowing boat in Port of Spain where he grew up as a white Trinidadian. He lived for the water, wore swimming trunks to school, was a Sea Scout and a champion rower. He even won a special royal commendation as a scout. It read: 'As a King's Scout you have prepared yourself for service to God and your fellow men, and have shown yourself to be a worthy member of the great SCOUT BROTHERHOOD.' And it was signed, 'George R.' – the King of England.

Jack was a tough child. As a baby he had fallen out of his

high chair onto a broken bottle. 'Anybody who called me "scarface" was in trouble fast, I can tell you.'

His father was not an educated man – more 'street smart' than anything else is the way Jack describes him now. The boy's first religious awareness came when he realized that he did not go to church – a Christian church. He had no religious education from his father. All he ever said was, 'Jews don't believe in Jesus.' He was taken to the synagogue on high holy days but it was usually in a makeshift building because at the time all the synagogues were temporary affairs. And he never had a Bar Mitzvah.

As with Oskar, the war came nearer as Jack grew into a boy. In his beloved rowing boat he watched the submarine nets go up outside the Port of Spain harbour. More and more British soldiers roamed the streets of Trinidad. The British national anthem, 'God Save the King', was heard more and more, and Jack learned the words so he could join in. He started to read newspapers and go to the cinema as the war came closer. The newsreels of those days had a lot about a 'comic madman' called Adolf Hitler. 'He was idiotic,' Yufe remembers being told, 'especially the hair-do.' His other main memory of the war years was that their neighbours – very pro-British – kept a picture of the battleship, HMS *Hood*, which was sunk by the Germans.

He moved to South America after the war, to Venezuela, to live for a time with his father's sister. His aunt was a Jew of course, but she was also a survivor of the concentration camps. And she was the only survivor in the family. Every other relative on Jack's father's side had died, either in the camps or in the towns where they had tried to set up home. Jack's grandparents, three other aunts and an uncle – all had perished.

The aunt also had more of an education than Jack's father and spoke several languages. To Jack she often spoke Yiddish, and he gradually picked it up. She was a strong, beautiful woman with a pale skin and thick, dark hair. But it was not an easy relationship between them. She had been too much scarred by the war – she had actually given birth in one of the concentration camps. But she had been alone, the child's father having been killed in the battle of Berlin. Like many of his generation, Jack came to understand the pain within his aunt, but he always kept his distance. When she suggested he go to Israel, Jack was torn. He was not sorry to get away from his aunt but he had no desire to go to the Middle East. The place, and what it stood for, had no attraction. His father, however, came back into the picture and reasserted his authority. He thought it would help Jack to see Israel.

The family had a couple of relatives living on the outskirts of Tel Aviv. In better circumstances they would have been quite unsuitable: they were Russian speakers, left-wingers, and their area was a scruffy, shanty neighbourhood away from the main town. But, protesting all the way, as he now puts it, Jack was sent to join them.

'I was soon homesick. Yiddish isn't Hebrew and I couldn't understand the language. Most of the music had that Middle-East Arabic twang, very strange. Israel was full of immigrants no better off than me and with just as little understanding of the place. Teeming, chaotic. After a few weeks I moved north to a kibbutz near Nazareth and managed to find work. But I was mad at my father for making me a mere labourer so far from home.

'Things gradually got a bit better, especially after I moved to another kibbutz on the Sea of Galilee. At least I could fish.'

The fishing style in Galilee was different from that in the West Indies. The catch was sardines, which could only be caught on moonless nights. 'I used to ride out into the middle of the lake in a sardine boat, towed by a motor launch. The engines would be shut off and everything became silent and black. You could see the black hulks of the Golan Heights and the only thing that broke the silence were occasional rifle shots. That's how I came to feel the Arabs as enemies. I couldn't see them, but I could feel their hostility.' Against that background he pulled in his fishing nets. It was all very different from the bright clear waters around Port of Spain but, as before, there was war across the water.

Later, Yufe did two and a half years in the Israeli navy, and even visited New York aboard the flagship *Misgav*. Then, in the early 1960s, he decided to emigrate to the USA (his father was then living in San Diego in California). On the way he stopped off in Germany to meet his brother.

The first coincidence was that Oskar was waiting at exactly the right point on the railway platform when Jack got down from the train. However, the meeting was not a success. They needed an interpreter but even so the encounter was painful. Jack was told by the interpreter that Oskar's stepfather, the man who had married the twins' mother on her return from the West Indies, did not know there were part-Jews in the family – and must never know.

Over the years Yufe has become a successful businessman. He has an American wife, two daughters and runs his own clothing store at Chula Vista in California. He wears a Star of David but does not regard himself as a very religious man. Most of his employees are Mexicans, and Yufe now speaks Spanish himself.

Once a year, around Chanukah (the Jewish holiday near Christmas), Oskar used to send Jack a Christmas card. Yufe's wife would send the Stohrs a card in return. Jack, the memory of that meeting on a German station still painful, could not bring himself to do so.

Oskar has two sons now, both raised as Catholics, and he still regards himself as a Catholic even though he does not agree with some of Rome's more rigid dogmas (such as the ban on contraception). He works in industry ('in a supervisory capacity') but will not discuss the exact nature of his job, nor disclose where he lives now. After all these years he still does not want it publicly known that his father was a Jew.

With such different backgrounds and their strong feelings about Judaism and its part in their lives, it is perhaps surprising that Oskar and Jack agreed to get together for the Minnesota study. But they did.

It was Jack who made the first move. In the summer of 1979 his wife read about the 'Jims'. Jack was interested, perhaps because the Minnesota study offered him the opportunity to meet Oskar on neutral territory.

Yufe wrote to Professor Bouchard and asked if the researcher could help them meet. 'This was my excuse for us to get together,' he admits now. 'I don't think we could have done it any other way.' Bouchard's study paid Oskar's expenses, and in November 1979 the reunion took place. It was an extraordinary meeting. Although they had not met properly for forty-six years (save for that uncomfortable encounter on a German railway station), uncanny coincidences were immediately apparent. Yufe, who still talks in the syncopated West Indian lilt, was the first to arrive in Minnesota, and went with Bouchard to the airport to meet his twin. He was curious to see whether his brother was 'another Jack Yufe – or completely different'. The first few seconds gave him the answer.

There was a man standing at the airline gate. Yufe originally had no idea what his brother would look like all these years later, but now he had no doubt.

'Oskar,' he shouted. The man turned – and Yufe gasped. Stohr was wearing wire-rimmed spectacles, rectangular but with rounded corners. They were exactly the same as Jack's. Stohr was also wearing a shirt with two breast pockets and epaulettes, as was Jack. Oskar's was light blue, Yufe's dark blue. Both men wore short, clipped moustaches, and their hair receded in exactly the same way. And that was just the beginning.

This second meeting of Oskar and Jack was more successful – though some matters were still unresolved. But, during their week in Minneapolis, they found they had all sorts of things in common.

Some of the coincidences were more 'scientific' than others. For instance, in filling out Bouchard's questionnaires, both found that they fell asleep easily in front of the television. Oskar absentmindedly stored rubber bands around his wrist – and so did Jack. Both found out during the week that they liked reading in restaurants, flushing the lavatory *before* they used it, reading magazines from back to front, and dipping buttered toast into their coffee.

Other coincidences became apparent only by accident as the week went by. On one of their nights off from Bouchard's study, for instance, they visited a hypnotist. The man was counting backwards, trying to induce a trance. 'Eight . . . seven . . . six . . . fi—' Oskar suddenly interrupted the proceedings with a loud sneeze: 'AATCHOO – ' 'He does that all the time,' said Oskar's wife. 'It's a little joke of his.' Yufe was stunned. For years he had never been able to enter a crowded lift without letting out just such a loud sneeze, in order to watch the reaction of the others.

The scientific results on Oskar and Jack, as with the other twins seen at Minnesota, have not yet appeared. But Bouchard has shown me a preliminary report on some of the psychiatric similarities and differences between the two men. In it Bouchard maintains professional confidentiality by disguising the identities of the twins, labelling them only as 'twin A' and 'twin B'. 'Twin A' has been an 'alcoholic' at some stage of his life, showing 'mild withdrawal symptoms and blackouts; when drinking he physically abused his wife'. On the other hand, twin B, who had once been a drinker, 'avoided alcohol because it made his narcolepsy worse'.

In contrast, both twins had 'recurrent anxiety' and both had explosive anger. 'Twin A', runs the report, 'was physically violent during his anger outbursts. . . . Twin B was never physically violent (he did not drink) but he was verbally abusive during anger outbursts to the extent that it caused him social embarrassment. Twin A was treated by psychotherapy during the preceding four years for an atypical depression with occasional suicidal thinking. His twin had no such problem.'

We can sum up Oskar and Jack by saying that there appear to be some extraordinary coincidences between them, in their behaviour, manners and habits, and that there are some similarities in their psychiatric profiles. There are, however, some important differences in these profiles also.

## Daphne and Barbara, the 'giggle twins'

These British twins became known in Minnesota as the 'giggle twins' because of the most notable feature they have in common: they both laugh 'more than anyone else we know'. Their voices are the same, and the way their bodies shake with laughter is identical. Both are very attractive, slightly mischievous, slightly overweight characters with distinctive teeth that give their faces an identical appearance when they laugh.

Daphne lives in Wakefield in Yorkshire, where she is a housewife, and Barbara's house is in Dover, where she is a clerk in local government. They were born in July 1939 in Hammersmith Hospital in London. Their mother, they discovered later, had been a Finnish student in London, and they were immediately put out for adoption. Adopted separately, they did not meet up again until May 1980. The most obvious difference between these two women is that Barbara wanted to find her twin but Daphne never felt that urge.

Daphne had known since she was eleven that she was a twin, but had done nothing about it. Barbara was twenty before she learned the truth. The immediate spur to the reunion came much later, however, when the twins were thirty-three. The couple who had adopted Barbara had both died while she was still at school and she had been taken in by an 'aunt'. It had never even occurred to her that she might be adopted, not until she needed a copy of her birth certificate when she applied to join a pension scheme.

That was how she learned that her real name was Gerda Barbara and that she was the daughter of a Finnish woman named Helena Jacobson. At first Barbara wanted to find her real mother. She wrote to a Finnish newspaper, and a journalist there helped trace the fact that Helena had returned to Finland soon after the twins' birth. But then she had married a German and went to live in Germany. Unhappily, Helena had committed suicide in 1943 by taking an overdose of sleeping pills.

Next Barbara turned her attention to her twin. Her birth certificate, in recording the fact of her arrival into the world, also gave the time. This is the way British birth certificates distinguish between twins. Barbara managed to trace the midwife present at their birth – miraculously she was still alive and even remembered the birth (twin births tend to be remembered by all concerned). The midwife's name was May Lambert, and she helped obtain a copy of Daphne's birth certificate. This showed that Daphne was born twelve minutes before Barbara and that Daphne's real name was Dagmar Margaret. May Lambert recalled that a couple had arrived to take Daphne away but could not remember who they were.

Barbara now tried advertising in newspapers. No luck. So next she tried going to law.

Britain's adoption laws are specifically designed to stop a

mother finding a child she has put up for adoption; it is generally felt that this is the best way of avoiding the heartaches and problems that might arise from a mother's 'second thoughts' later in life. But this is essentially what Barbara wanted: she wanted the courts to open up Daphne's adoption papers to her.

Barbara asked specialist lawyers for advice and eventually took her case to Westminster County Court in London in an attempt to get the Registrar General to release Daphne's papers. It was an extraordinary piece of bravado because she chose to plead her case herself. She is normally very shy, not especially articulate and scarcely familiar with the law. But to get her way she had to address the whole court and face a formidable judge.

It worked. In straightforward terms she told the judge that it was important to her to find her twin, that it was not really like a mother seeking to reclaim a child she had already given up. The Registrar General would not give up Daphne's papers, but he did agree to trace her and ask if she would be prepared to meet her twin sister.

Daphne's parents, on receiving the letter from the Registrar General, had been unable to make up their minds whether to tell her Barbara was trying to get in touch – so they had asked her husband, Peter. He had taken matters into his own hands and shown the letter to his wife. Daphne immediately wrote to Barbara, suggesting that they meet at London's King's Cross railway station.

It was a curious reunion: they were strangers, in a way, though they did not feel it. But they did not embrace or shake hands or stand and look. What happened was that each noticed the other had a crooked little finger. This family defect meant that both of them are unable to play the piano properly. Immediately, they began comparing their fingers on the station platform.

The list of intriguing similarities between Barbara and Daphne is, if anything, longer than Oskar and Jack's and even more difficult to measure scientifically. For instance, before their reunion:

Both tinted their greying hair with the same shade of
auburn

Both enjoyed the novels of Alistair MacLean and Catherine
Cookson

Both used to read *My Weekly* (a woman's magazine) and
then stopped

Both clutch the banister for fear of falling downstairs

Both met their future husbands at town hall dances when
they were sixteen and both were married in their early
twenties in big autumn weddings, with all the trimmings
including a choir

Both worked in local government as did both their
husbands when they met them

Both laugh more than anyone else they know

Both describe their husbands as quiet, hard-working men

Both suffered miscarriages with their first babies; then each
had two boys followed by a girl. (Barbara and husband
Frank, manager of a handyman's shop, stopped there but
Daphne and Peter, a computer salesman, had another son
and daughter)

Both are careful about money

Both have the same favourite colour – blue

Both hated games and maths at school

Both had fallen downstairs aged fifteen, an accident that
had left them both with weak ankles

Both had been Girl Guides

Both had taken ballroom-dancing lessons

Barbara was evacuated to Silchester during the Second
World War; Daphne moved there later

Barbara's maiden name was Sandal. Daphne lives in Sandal

Both have a history of fad diets and put on weight very
easily

Both drink their coffee black and cold (no sugar)

Both love chocolate and liqueurs

Both hate heights and are squeamish about blood

Both have no sense of direction

Both arrived at their reunion meeting (at King's Cross)
wearing a beige dress and a brown velvet jacket

Both have a habit of pushing up their noses, which both call
'squidging'

Both gesticulate wildly and often catch themselves, say to
   their hands 'keep still, won't you' and sit on them
Both spread their left hand over one side of their face when
   they are nervous
Both use the same expressions when they tell off their
   children

This list is already quite long but it does not stop there. We
know a little bit more about Barbara and Daphne in Minnesota than we know about Oskar and Jack. We know for a
start that they were one of the liveliest pairs of twins to visit
Bouchard. We know that they completed more tests in their
week at Minneapolis than any other set of twins. And we
know that they invented a cocktail at their hotel during their
week there. (They called it 'Twin Sin': 1 oz vodka, 1 oz blue
curaçao, 1 oz crème de cacao, and 1½ oz cream.)

Barbara and Daphne came out very similar on Bouchard's
tests although they were brought up in different social classes.
Barbara was adopted by a municipal park gardener and his
wife in the London borough of Hammersmith, whereas
Daphne grew up in the more prosperous town of Luton as
the daughter of a metallurgist who worked for Vauxhall
Motors (the British subsidiary of General Motors). Yet the
pair scored almost identically on the vocabulary test; if Barbara did suffer any deprivation by being brought up in a less
'verbal' background, she has caught up now.

The medical tests showed that, although both twins are
within a fraction of 5 foot 3 inches, Daphne is 20 pounds
lighter because she diets. Both were found at Minnesota to
have a slight heart murmur and slightly enlarged thyroid
glands – though in neither case is it serious. They also have
matching brainwave patterns, responding almost identically
both to loud noises and soft music.

Bouchard was particularly intrigued to find that despite the
class difference both women became silent in serious discussion. Neither of them could be drawn on controversial topics
– Rhodesia, Northern Ireland, women's lib. Both liked the
Queen but even then would not argue for or against the
monarchy. In fact, Bouchard found that neither woman had
ever voted, on the grounds that they did not know enough

about the issues, *except* once when they had *both* worked as polling clerks.

At the end of their week in Minneapolis, as they settled down to a spectacular lunch to celebrate the end of their week-long testing, one final coincidence emerged. Both had told Bouchard the same lie. 'We both said we wanted to be opera singers and neither of us can sing a note,' Barbara confessed. And they both broke into peals of laughter, yet again.

## Jake and Keith, 'blood brothers'

The first coincidence in the lives of Jake Hellbach and Keith Heitzman occurred without either of them knowing it. They were born in Algiers, a suburb of New Orleans, and adopted into different families. But, as luck would have it, the entry for Hellbach in the New Orleans phone book was in the next column to, and right opposite, that for Heitzman.

These twins are known in Minnesota for the fact that, to Bouchard, they were more alike than any of the others. 'Some identical twins, after all the years apart, don't look all that identical . . . but these do.' Both are tall, dark and handsome men, slim and given to slightly sardonic smiles which spread only gradually over their faces. They resemble each other so much in fact that, during their week at Minnesota, Bouchard actually mistook Keith for Jake over lunch on the Tuesday, the only time this has ever happened. 'It's not only the physical similarity,' says Bouchard, 'but their similarity in being so quiet is also striking. They even sit with their arms folded in the same way.'

After they were separated, Jake and Keith grew up on different sides of the Mississippi in Jefferson County, Louisiana, just outside New Orleans. They attended rival high schools. Jake eventually settled in Abita Springs and Keith in Metairie. Jake is now a pump mechanic and Keith a welder.

The chain of events which led to their reunion began a long time ago. While they were growing up, neither knew he was a twin – their adoptive parents judged it best not to tell them. But Keith had always believed that he had a brother. His feeling was confirmed when, in June 1979, he saw his birth certificate and found out that he had been the first-born of a set of twin boys. Then he started looking for his brother in earnest.

'I was never too bothered to look for my natural parents,' he now says, 'but I did want to find my brother.' The next move came when his adoptive mother contacted Bouchard. She had read about the Minnesota study in a magazine, and she asked the professor to help find Keith's twin. Bouchard put one of his assistants on to the problem but their inquiries came to nothing. As a last resort, Keith paid for his own advertisement in the New Orleans afternoon paper, the *States-Item*. That did the trick.

After the ad appeared, the editors of the paper became interested in Keith's search and wrote him up in an article. Jake Hellbach's adoptive mother had not seen Keith's ad, but she did see the article. That was when she told Jake that she realized Heitzman must be his twin.

Since Keith was already in touch with Bouchard it was not long before he and his brother were on their way to Minneapolis. Although they were the most similar twins Bouchard had yet encountered in terms of their bearing, manners and habits, they did show some scientific differences. For a start, Jake is left-handed, Keith right-handed. This affected how they did some tests. For instance, in one test they had to draw a house: Jake put the door on the left, Keith on the right. Otherwise the houses were very similar. It is possible that they may be mirror-image twins, identical but with certain key attributes – the crown of the hair, fingerprints, whorls and so on – reversed. Bouchard is not certain yet, since they were the first pair he had seen who had differed in this way. Another difference is that Jake has a bigger head though a slightly smaller body.

Like the other twins, Jake and Keith made several intriguing discoveries about themselves at Minneapolis, among them:

Both are allergic to ragweed and dust
Both did poorly in school
Both avoided gym classes in school but enjoyed art
Both have a strong interest in guns and hunting

Both reload their cartridge shells in the same way
Both have an unquenchable appetite for chocolate and
    other sweet things
Both dress in a similar manner, and are especially fond of
    large cowboy hats
Both 'put off until tomorrow what should be done today'

These two also show perfectly what happens to many identical twins who meet up after years of separation. Although it was Keith who did the looking because he always felt he had a brother, Jake had always wanted a brother. 'Finding a blood brother as an adult is great – to say the least,' he says now.

'They're making up for lost time,' says Stacey Heitzman, Keith's wife. 'There's just no separating them now.' 'They just think the same,' says Caroll Heitzman, Keith's adoptive mother. 'They are both attuned to what each other is thinking . . . being separated you wouldn't imagine they would have this.' And, Stacey adds, 'the twins can communicate using only partial sentences no one else understands.'

Outside the laboratories, when they were in Minneapolis, Keith and Jake were still a source of interest. On one occasion, unbeknown to each other, they entered adjacent phone booths in their hotel simultaneously – and called each other. After spending their days separated, for the purposes of the testing, they would hurry back to each other in the evening and the normally quiet, reserved men would, in each other's company, become comics who, according to Stacey, 'act like they've been practising as a team for years'. While they were in Minnesota, a CBS crew was filming the experiment for a TV programme. 'Today CBS,' quipped Jake, 'tomorrow Johnny Carson.'

Both had also started keeping a scrapbook since they had been reunited. These books contain news clippings about their reunion, their birth certificates, snapshots of their first birthday spent together, on 4 January 1980. But the pair have also received a large number of letters from strangers who seem moved by their story, showing the special feeling the bond between twins evokes in many people. 'I really don't have a specific reason for writing,' said one fifteen-year-old Metairie girl in her letter. 'All I know is that, when I read about you finding your brother, I felt so happy inside that I cried.'

## The spooky diaries of Dorothy and Bridget

Dorothy Lowe and Bridget Harrison share the single most perplexing coincidence in this entire book. Like other twins in the Bouchard study, Bridget and Dorothy were separated only weeks after their birth in 1945 and did not meet up again until they were thirty-four in 1979. During the intervening years neither woman knew she was a twin *and yet* . . . for the year 1960 (and only for that year) both women – then girls of fifteen – kept a diary. Moreover, they bought the *same* make of book, the *same* type and the *same* colour. The entries were not that similar but the days they filled in and the days they left blank *did* correspond.

Dorothy and Bridget were the first British twins to take part in Bouchard's study. Like Daphne and Barbara, they were adopted by parents of different social classes. Yet the only medical and psychological difference seems to be that the twin raised in the more modest circumstances has distinctly worse teeth. That apart, however, Dorothy and Bridget must number some of the most extraordinary coincidences of any pair of twins.

When they arrived (together) at Minneapolis, Bouchard had another of those shocks he has come to associate with the airport. Both Dorothy and Bridget had seven rings on their hands; they also wore two bracelets on one wrist and a watch and a bracelet on the other. Coincidence? Or a good example of the infinitely complex interaction between the genes the two women have in common and their environment? At the back of Bouchard's mind is reasoning that goes something like this:

Among the genes that the women have in common is one which gives both of them long, slender hands; on top of this they are both attractive and have made successful marriages to fairly wealthy men; which means they can indulge their out-going personalities, which implies a liking for nice clothes and jewellery. Combined, these genes would make it likely that they would adorn their hands with rings and bracelets of some sort. The exact number and arrangement of jewellery was then a coincidence on top of this genetic and environmental interaction. That, anyway, is the kind of explanation Bouchard and his colleagues prefer, at the moment, to telepathy or ESP or astrology.

Before the week was out, however, they had shown many more coincidences which are much more difficult to explain away. For instance:

Both women took piano studies to the same grade, then stopped after the same exam

Both had meningitis

Both had cats called Tiger

Both collect soft, cuddly toys (both gave each other teddy bears at their reunion)

Both are avid readers of historical novels, Dorothy of Catherine Cookson, Bridget of Caroline Marchant (which is Catherine Cookson's other pen name)

Dorothy had named her son Richard Andrew, and Bridget had called hers Andrew Richard. Dorothy's daughter was called Catherine Louise and Bridget's Karen Louise (Bridget actually wanted to call her daughter Catherine but changed it to Karen to please a relative)

Both leave their bedroom doors slightly ajar

Both wore almost identical wedding dresses and carried the same flowers

Both have the same favourite perfume

Both are anxious about their legs being too thin

Both call themselves short-tempered, strict with their children, impulsive

Both liked hockey and netball at school

Both have the same make of washing machine

Both describe themselves as 'snobby'

During their week, Bouchard also found that they sat in the same way, and that many of their gestures were identical, too. For instance, Dorothy and Bridget both wave as they speak, covering their mouth with the right hand if they are even slightly nervous about a subject. They also both push back the cuticles of their fingernails while they are talking.

Standing, they cross their arms in the same way and cross their legs with the same leg slightly in front.

At Minneapolis they also found that, although Dorothy smokes and Bridget does not, their lungs were in good shape. Their respiratory reaction to stress – after running, for instance – was just the same. Dorothy and Bridget were not the only twins where one was a smoker and the other was not, nor the only case where the smoking twin's lungs were no worse than the other's. This could be important: the indirect evidence so far, which is based on very few cases, begins to suggest that *some* people (it may be very few) are constitutionally unlikely to suffer the adverse effects of smoking, including perhaps cancer.

Bouchard's colleagues also investigated the psychiatric history of Dorothy and Bridget and this, too, is interesting. 'Both twins,' the report reads, 'tended to be moody.' Both were anxious and they were tested on a variety of medications. Several were effective but the best for both of them was the same one, Propranol. In the stress experiments, the psychiatrists found that both the twins tended to respond with physiological symptoms – mainly stomach upsets, shortness of breath and a pumping heart. In fact, both actually became ill when they had to take the Minneapolis tests. They almost always felt sick – and in some cases they actually vomited. One twin developed the symptoms sooner and they lasted longer. This twin responded with hyperventilation (short, sharp breaths) during some of the medical procedures, while the other twin 'experienced light-headedness and a feeling of faintness'. Both, it seems, had a history of breathlessness as well as a 'tingling' in their extremities during moments of stress. 'Overall,' concludes the report, 'it appeared that twin A tended to react more intensely, both physiologically and emotionally, to stress. Environmental variation may account for this difference in that twin B was raised in an environment which required that she develop some "toughness" to survive. She also disliked the extreme responses of twin A, saying she certainly didn't want to be like her.'

Since their reunion the coincidences between Dorothy and Bridget have continued. When I went to meet them, Bridget admitted that, had her sister not phoned beforehand, she would have been wearing the same skirt as Dorothy. Another example: they agreed on the same present for each other for their last birthday. Dorothy was tempted to buy something quite different she saw in a shop – but decided in the end to stick with the agreed choice. Bridget, tempted just the same way, gave in and bought her the very thing that tempted Dorothy.

When they met, Dorothy and Bridget had no difficulty talking to each other. For the first couple of months their phone bills were huge (Leicester and Burnley are 100 miles apart), and they felt that they were taking over each other's lives. But things then 'settled down'. Bridget's husband Mike seems to have coped better with having a ready-made new family thrust upon him. Dorothy is, if anything, more excited by the reunion than Bridget, who is more wary of the disruption it might cause in the lives of their families. On the other hand, Dorothy was lonely as a child whereas Bridget had an adoptive sister. But they both say they have gained in confidence since finding their sister and that they feel more secure.

## Irene Reid and Jeanette Hamilton

These twins were born on 2 April 1944 and separated when they were about six weeks old. Irene found out that she was a twin when she was sixteen and wanted to get engaged. She gradually began to think more and more about her sister. She had planned to start looking when she was twenty-one, but she had a very unhappy marriage and put off the search several times. Eventually, she became divorced, but it was not until she was thirty-three, in 1977, that she finally decided to do something about it.

She watched a programme on British television about the reunion of Bridget and Dorothy, but even then she was in two minds. Everything they said about the joy of meeting up, she felt; but she did not want to hurt her adoptive father's feelings. (And Bridget had tried to find Dorothy for seven long years – Irene did not want to go through that.)

She went home to Scotland (she lives in England now) and checked her birth certificate. She found she had been born Agnes Grey, and that her sister was called Barbara.

Eventually she contacted the *Sunday Post*, a large-circulation Sunday paper in Scotland, and the paper ran a story about her search for her twin sister.

On the Sunday the story appeared Irene was at home in Yorkshire, outside the area of distribution of the *Sunday Post*. About 1.40 in the afternoon the phone rang. At first she could not decipher the name of the caller. Then she realized that the woman was saying she was Barbara Grey – it was Irene's twin. They cried for fifteen minutes.

Over the phone, when they could talk, they began to compare notes, and found they both used the same type of hair dye (amazingly enough, that was what they talked about during their first conversation for thirty-two years). They met two days later.

It turned out that Barbara's (Jeanette's) mother had seen the article and phoned her daughter. 'Oh Mum, oh Mum,' Jeanette had said, 'what should I do?'

'I wanted this to happen,' replied her mother.

Later, it turned out that Jeanette's young baby (who was just eighteen months old) had been crawling over the copy of the *Sunday Post* when he had seen the pictures on the page, the pictures with Irene's face in them. He was barely able to talk but he did manage to say: 'Mummy.'

Later, when the eighteen month old was introduced to Irene, and had been suitably confused by the likeness (he has said: 'I have two mummies'), he started to call her 'Aunty Mummy.'

When they met on the Tuesday Irene had her house sparkling. The weather was very bad; Jeanette was late and flustered when she arrived. Before they could take it all in, or cuddle in the hall, Jeanette said: 'It's me – but I haven't got a blouse like that!'

And their similarities? Well, they are numerous, to say the least:

Both love ice skating
Both love crocheting and sewing
Both talk with their hands. In their first meeting Ian,
    Jeanette's husband, would often remark when Irene made
    some movement, 'That's just what Jeanette does'

They both have marks on their bodies in the same place,
where they had accidents as children
Both got the marks when they fell off fences
Irene has always had a pain in her right leg, from her hip
down to her knee. There is no reason for it but . . .
Jeanette has a wasted muscle in her right hip. She never
gets the pain, although she has the disease
Both have two children conceived at much the same time
When they were in Minneapolis they both had a 'psychic'
experience. One day Irene was with the psychiatrist when
she got the feeling that Jeanette was in trouble. Half an
hour later, Bouchard called Irene, she says, to tell her
that Jeanette was indeed poorly. Two days later the same
thing happened but the other way round; this time Irene
was ill and Jeanette sensed it
Both fold their clothes in the same way; when they are
putting things away both button up every other button on
their blouses and shirts
Both have the same favourite colour: blue
Both have the same favourite perfumes: Revlon Intimate
and Estée Lauder
Both are mad about Turkish delight and liquorice allsorts
Both ad-libbed a lot in response to the Minneapolis
questions. One question reads: 'Would you tolerate
another person's religion?' Independently each scored out
'tolerate' and replaced it with 'respect'. Similarly, in
response to the question: 'Do you like to be dressed in
today's fashion?' both replied: 'If I've got the money'
Both took the same amount of time on the test
Their IQs had one point between them
Apparently Bouchard felt their husbands look alike but the
twins do not agree with this
However, they do think their friends look alike
Both think their adoptive fathers resembled each other
Both count the numbers of wheels on a lorry when it
passes, and have other similar obsessive habits

Both twins were overwhelmed by the similarities. 'If I had
passed her on the street,' says Irene of Jeanette, 'I would
have known, I would have known.'

Then there are the psychiatric similarities. Both wet their beds as children, one until she was six, the other until she was ten. Both were treated with a mild tranquillizer for nervousness and tension during a time when both their marriages were under stress (in both cases their husbands were drinking heavily).

Both women were terrified of going into the sea above knee level. They both liked to paddle and had, independently, developed the same way of approaching the sea: they would tread carefully *backwards* into the water until it came up to their knees – a curious habit since they would be more likely to fall over.

Both also hated to enter the booth for the brainwave tests (they christened Dr Lykken 'Burl Ives'). And both hated heights: when they went to Minneapolis it was the first time either had flown in an aeroplane and when Dr Bouchard took them to a multistorey car park both had to rest for half an hour before they could calm down enough to go shopping.

But we must not overlook the differences. Irene had a much unhappier marriage than Jeanette and is now divorced. One had two boys, the other two girls. Jeanette likes to sing, Irene to dance. When they were asked to draw a man at Minneapolis Irene drew a matchstick man, Jeanette a fat, rounded snowman, almost the opposite. On the flight to Minneapolis they were offered the opportunity to visit the flight deck: Irene went up but Jeanette refused – she was too terrified. They feel that Irene is a much quicker person, much more likely to make a blunder, whereas Jeanette is slower, more reflective, perhaps because she has a happy marriage. (Irene is a night nurse, still training.) Jeanette is quite good at maths and enjoys it – Irene on the other hand, and despite her nurse's training, cannot stand maths. And Jeanette has a remarkable memory – right back to her time in the pram. Yet she never remembered she was a twin.

## Terry Connolly and Margaret Richardson

One point needs to be borne in mind in considering this pair of twins. After they had been to Minneapolis, all the British twins got together for a grand reunion in the autumn of 1980.

By and large the other British twins felt that Margaret and Terry were less alike than any of the others. Yet they always dressed alike when attending functions together, and dressed identically when they visited Bouchard. And they would address each other as 'Hey, twin,' for example, rather than by Christian name. This seems to be an instance of twins actually seeking to create similarities, 'twinning', as Susan Farber calls it.

Margaret and Terry were born on St Valentine's Day, 1943, in Leicester. Margaret was adopted when she was six weeks old but Terry had to wait until she was eight months before a family came forward for her. The name of the family was Quick.

Each twin remained completely unaware of the other until 1978, when Margaret's father told her that she was a twin and that her sister had been adopted into a family with the unusual name of Quick. Straight away Margaret tried calling all the Quicks in Leicester (there were only four in the phone book). First she found an 'uncle' of Terry's but, before very long, she got on to Terry herself. 'It was,' says Terry, 'the most emotional thing that has ever happened to me.' In fact she was in bed, ill, for weeks. She had all the symptoms of flu without the cough and colds. But she had to stay in bed.

Several similarities were spotted early on:

Both were married on the same day of the same year, at
   roughly the same time
Both like the same food: red peppers and courgettes
Both had wanted and planned to have four children
The children were conceived and born at roughly the same
   time

| *Margaret* | *Terry* |
| --- | --- |
| Luke 11½ | Beth 11 |
| Ruth 9 | Tim 10 |
| Ben 4 | Matt 5 |
| Jim 3 | Meg 3 |

Terry had wanted to name her first daughter Ruth, just like
   Margaret, but had been dissuaded because unpleasant
   neighbours had a daughter of this name

Both were claustrophobic at Minneapolis
Both giggled a lot, happy to be away from all their children
    at last
Their children are similar in some respects. Tim and Luke,
    for example, are very shy and withdrawn and physically
    very similar
Both use the same perfume – Boots No. 7 – but then, as
    Terry says, it suits the dry skin they both have

Now let us look at the differences. Margaret has three boys
and a girl whereas Terry has two of each. Terry says she
married an extravert engineer, the managing director of his
firm, whereas Margaret married a much more 'negative' man,
the warden of a teachers' centre in Nottingham. Margaret
had polio as a child, which left her with a gammy leg, so she
rarely wears skirts, preferring trousers almost all the time.
Terry is often in trousers too, but she does not have the same
aversion to skirts.

Margaret went to a private school whereas Terry went to
a grammar school (both passed the eleven-plus). Both got
seven O levels and took up teacher-training, yet Terry took
English and PE whereas Margaret stuck to general studies so
that she could teach young children. Margaret's parents
adopted another girl, so she grew up with a sister, whereas
Terry's parents could never afford an addition to the family.

Neither twin feels any 'psychic bond' between them. They
feel that there are some similarities and coincidences between
them, but, as Terry Connolly puts it: 'Nothing more than
what any two people pulled in off the street would show. Too
much is made of all these coincidences.' On the other hand,
she had made a list of similarities, and called me back to read
it out to me.

## Ethel and Helen, and their 'psychic bond'

Ethel Gaherty and Helen Moscardini differ in one important
respect from the other twins in the Minnesota study: they
were not separated immediately after birth. They grew up
together until they were two and a half, then were adopted
into different families. Everyone tried to make them forget
they were twins, but the memories stayed.

Ethel Gaherty always knew there was another little girl, somewhere, just like her. Now well into her fifties, she can still remember playing with her sister, licking a pastry dish clean together, and the fact that her twin had had an illness that nearly killed her.

'In the days before the war,' says Ethel, 'adoption agencies and adoptive parents tried to erase the past. I suppose they meant well: they gave us new birth certificates with our new names on them. But, like me, my sister couldn't forget.' The twins grew up with a yearning to find each other and be together. 'It was a compulsion really,' says Ethel rather wistfully.

The girls were born on 2 March 1925 at the Herman Kiefer Hospital in Detroit. All went well initially but then, towards the end of 1927, their mother fell ill and died. That was how their adoption into different families came about. Although they did not know it, both Ethel and Helen grew up in Michigan, not far from each other. The first unusual coincidence occurred when they were eight or nine.

As Ethel tells the story: 'I was at a picnic – a church picnic it must have been – because all of a sudden a visiting priest came up to me and said, "Why Helen, what are you doing so far from home?" Then, of course, he found out I wasn't Helen and we all had a good laugh. But he did say that I looked enough like her to be her twin.'

At last. 'I had always been pressured not to talk about it, but I'd always known I'd had a twin. I managed to persuade the priest to promise he would get this "Helen" and me together.'

The priest was as good as his word, and now came the most extraordinary coincidence of all. Ethel wore her favourite dress for the occasion (it was organdie with a stiff bodice and flared skirt); Helen (who, of course, lived in another town) borrowed a new dress from a friend so she could look extra smart for the meeting. They were *identical* dresses. More than that, they had brushed their hair in the same way.

Throughout their teens the girls kept in touch by letter though they managed to see each other only every two or three years. When they were eighteen Helen left her adoptive home and went to work in Chicago. Ethel stayed on at college until, at twenty-one, she graduated as a nurse and moved to Chicago as well, in order to be with Helen.

What was extraordinary was that until this point the girls could not be certain they were twins. Throughout the 1920s, 1930s and 1940s the adoption agencies were very strict about adoption documents, and the girls could not obtain their original birth certificates. But then Helen had a brainwave: she told Chicago's famous mayor, Richard Daley, the details of their story. She said they were convinced they were twins and described the coincidence over the dress. Within twenty-four hours the mayor had found their birth certificates. The certificates proved what they already knew, but says Gaherty: 'It gave us a feeling of security. I belong to her, she belongs to me.'

When Ethel and Helen arrived at Minnesota, they had not been only recently reunited, like all the other twins, but had been seeing each other more or less regularly since 1946. Would they be more alike than the other twins as a result?

Helen has been a chain-smoker for thirty-eight years whereas Ethel has never smoked. Nevertheless, just as with Dorothy Lowe and Bridget Harrison, their heart and lung tests came out nearly identical. Years of smoking appeared to have had no effect on Helen Moscardini's lungs.

The other main difference between them cropped up in the psychiatric report. Twin B 'had developed a depressive neurosis since menopause' while twin A was 'emotionally stable.' Since even neuroses (like more serious forms of mental illness) may have a genetic cause, this finding could be important.

Recent anecdotal coincidences in the lives of Ethel and Helen mean rather less than in other cases, of course, because they have been in touch so much. Still, there is no shortage of them. Ethel loves to tell the story which illustrates what she calls a 'psychic bond' between them. 'One time I called Helen and asked "What happened to Rosa [Helen's mother-in-law]?" She had died twenty minutes before.'

Among the other similarities are:

Both Ethel and Helen have the same eye problems
Both have a fear of heights

Both have the same pattern of brainwave activity in
response to light and sound
Both have the same favourite colours – pastel pinks and
blues
Both always wanted to be nurses, and Ethel became one.
However, Helen had been through fourteen foster homes
by the time she should have started her training; she
wanted to get out into the world and be independent, so
she turned down her foster parents' offer to send her to
nursing college

Ethel and Helen are very conscious of their differences also.
They believe, for instance, that they have very different personalities, Ethel being the more outgoing. Their history of
illnesses has not been very similar, either. When I spoke to
Ethel she was in bed with a bad case of pneumonia, which
she developed after a few days at the tables at Las Vegas.
Helen had been there as well but she was perfectly well. They
think they are 'psychic' but only in so far as each other is
concerned and only in so far as they tend to phone each other
at the same time. When I asked Ethel what she thought Helen
was thinking at that moment she did not have any idea.

But they have strong views on the separation of twins. 'It's
a grave injustice to separate twins,' says Ethel. 'I think that
if you started out together you were meant to be together.'

## Gladys and Goldie, the 'love–hate' twins

Gladys Lloyd lives in Arizona, Goldie Michael in Michigan,
and they do not get on. Gladys and Goldie are continually
quarrelling, though they always seem to make up. They are
now fifty-seven, were separated at about six weeks and, says
Bouchard, 'they have a love–hate relationship. They carry on
all the time but they need each other.'

When they arrived (separately) in Minneapolis, each tried
to enlist Bouchard as an ally against the other, telling him
what they 'thought he should know' about their twin. He
tried to resist but before the week was half over he had to
take them to one side and explain that their animosity was
spoiling the experiment. They patched it up, partially at least,

until the end of the week and they still keep in touch. But their relationship remains uneasy.

In 1964, Gladys was married to a businessman who, behind her back, also wanted her twin as his mistress. He asked Goldie if he could buy a house for her, and a car, in return for her secret favours. It is not quite clear for how long this arrangement went on, but within a few weeks or months Gladys found out and divorced the man. Since then the relationship between the sisters has been off and on. When they were both pregnant, they were close. They were so alike then that Gladys's husband (another one) kissed Goldie by mistake as he came into the house one day from the office.

Gladys freely confesses, unlike all the other twins in Bouchard's study, 'I don't care for my sister; I hope I'm not like her.' Perhaps that only makes their list of coincidences all the more remarkable:

Both had nightmares as children, which they describe in a
    similar way (they imagine doorknobs and fishhooks in
    their mouths, and feel they are smothering to death)
In both cases the nightmares started in their teens and only
    stopped in the last ten to twelve years
Both developed diabetes at the same time
Both were bedwetters until aged twelve or thirteen
Both had the same school history
Both had been married *five* times
Both had the same phobias
Both went through a prolonged period of heavy drinking
    between the ages of fifteen and thirty (although neither
    Gladys nor Goldie would have been officially described as
    alcoholic)
Both, however, would have been regarded as drug abusers.
    Both began to use amphetamines for weight control when
    they were aged between thirty and forty. Gladys says she
    sometimes used as many as twenty pills a day. Goldie was
    admitted to hospital twice with a paranoid psychosis
    which the doctors believe was 'most likely' linked to her
    abuse of amphetamines

Both worked as barmaids and waitresses
Both were pregnant at the same time, with both their
   children. Their sons were born thirteen days apart
Both took to fixing up old properties, houses in Gladys's
   case, bars in Goldie's
Both preferred blue and pink
As girls, both liked horses
As children, both dropped their mother's pies and were
   spanked for it
Both had measles, chicken pox and whooping cough

Now they emphasize their differences. Gladys is better off, lives in a bigger house and drives a Cadillac, whereas Goldie has 'an older car'. They speak to each other no more than three or four times a year and then only through an adoptive sister, Betty. And although Gladys is quite proud of her 'psychic' powers she specifically excludes sister Goldie from these. Gladys was treated by a Beverly Hills psychiatrist for two and a half years, during which time, she says, she rather spookily predicted two people's deaths (the psychiatrist was impressed anyway). But Gladys is adamant she is not psychic where her twin is concerned.

No doubt the mistress business has something to do with the animosity between Gladys and Goldie. But perhaps hostility continues because they do not like the part of themselves that they see in the other. That may also explain their instability and drug abuse.

## Dan Sivolella and Michael Meredith

These twins have very few characteristics in common. They tend to the view that the whole Minneapolis project is an exaggeration, that there are not many coincidences between them and that the differences outweigh the similarities; in short, that they are no more alike than any other two people picked at random.

They were born on 11 November 1948 and were separated at six to eight weeks. They first met up again when they were seventeen and then again at nineteen. They are now thirty-three years old so have had plenty of opportunity to meet and influence one another.

When you talk to these twins you are struck most by their laconic attitude: they see nothing extraordinary in themselves. They met like this. Both twins had adoptive cousins. Both these cousins, as it happened, worked on a hot-dog stand in Evansville in Indiana. One night these cousins got to talking, and started on about their families. Next night they both brought along photographs of their relatives and spotted the striking similarity between Dan and Mike. The twins were laconic even then; it took them another three months to get together. When they did so they found:

Both like bowling and pool
Both had similar smoking and drinking habits ('bad habits'
   as they call them)
Both were involved in a vehicle accident at the same time
   (1965) and suffered similar minor injuries
Both were allergic as schoolboys, Dan to plants while Mike
   suffered badly from asthma
Both had their tonsils out as boys
Both have planters' warts on their feet, big calluses that
   have to be treated with medication to stop them being
   painful
Both are good eaters and gobble down everything that is
   put before them

But some of these points are not very specific characteristics, for the truth is that the differences between Dan and Mike are every bit as interesting as the similarities.

For instance, Mike stayed back a grade in school. The reason is not certain but it seems clear that they are different intellectually. Both have been married. Dan is married to his second wife; Mike has only just divorced his first wife and, although he plans to remarry (to a girl who, according to Dan, looks like his wife), at present he is single.

Dan has a boy of three and a girl of two, whereas Mike has two boys of seven and five. They dress differently, though they are more similar now that they are living closer together in Indiana (Dan used to live in California). Physically, they used to be exactly the same but, although their height remains unchanged, Dan has lost weight in his current job, which is in the same factory as Mike but is of a more manual nature. He is now the smaller by two waist sizes.

Dan and Mike are a good example of twins who, in truth, are not so close to one another.

This completes the number of identical twins raised separately whom Bouchard has seen and whose names he is willing to release. For a variety of reasons the other five sets of twins he has seen do not want any personal publicity, and two I now mention are therefore identified only by a 'case number'.

*Case history 1:*

*Twenty-three-year-old male twins, separated at about five days. Evaluated within six months after they had accidentally found each other*

These two homosexual men now live together and have a sexual relationship with each other. The incest taboo, even in these unusual circumstances, lacks any force for them, perhaps because they did not grow up together.

Their birth records list no abnormalities. At birth, the twins were equal in length though twin A was a little heavier (and taller at the time of their visit to Minnesota). They were separated after a few days, for which no reason is given in the records, and adopted soon after. There are remarkable similarities in their developmental histories.

Both boys had a fear of heights at an early age (from four until they were twelve). Both refused to climb walls and cried when placed on ledges. This fear gradually lessened, and by adult life each retained only a mild height phobia. One of the boys (twin B) wet his bed until he was seven, but this did not happen with his brother. Both had speech problems at an early age. 'These were described as difficulty pronouncing particular letters and syllables,' says the report, though the mothers could not remember which letters so we cannot tell whether these were the same. Early on in school both received speech therapy so that their problem was sorted out by the time they were about nine.

According to the records both boys began to show

hyperactivity when they were aged eighteen months to two years. The symptoms were a decreased attention span, distractability and an inability to sit still. Both children were seen by psychiatrists because the symptoms were so serious. Twin B's parents were told he had minimal brain damage although a neurological examination could find nothing. Nevertheless, he was given the drug chlorpromazine from six until he was fourteen in an effort to treat his hyperactivity, and again from eighteen to twenty. The other twin was never actually diagnosed as having minimal brain damage and took no medicines. Both had to attend special classes in school to remedy their learning difficulties.

Both twins were also prone to outbursts of temper as children: they would cry and scream without any provocation. Again twin B appears to have been more severely affected, so that he actually became physically violent and threw things. Moreover, he was admitted to hospital when he was quite old – aged twenty – for temper outbursts associated with physical violence. Records show that the outbursts were associated with hostile feelings towards his family, especially his father; there were hostile feelings against neighbours as well. On several occasions twin B stole things, but it always seemed to be an impulsive rather than a premeditated crime. He was put on probation for stealing when he was eighteen.

Twin A had no such record, but both boys did have a problem holding down steady jobs. Perhaps for this reason both remained relatively dependent on their families and were fairly anxious people in their late teens. Both confessed to recurrent suicidal feelings.

The report does not go into much detail about their sexual development. It notes that both are homosexual but adds that there are differences between them. Twin B, for example, has had many more partners.

The twins met by accident when they began to go to the same gay bar in a north-eastern town in the United States. Friends mistook them for each other, and it soon became apparent that the boys were twins. They met and, like so many twins in Bouchard's study, took to each other straight away. So much so that by the time they travelled to Minneapolis, they had started living together.

These twins shared several psychiatric symptoms but one of them, in this case the second-born and the shorter (twin B), was the more severely disturbed. Could this difference between twins A and B tell us something about the way environment can affect genetic differences?

*Case history 2:*
*Sixteen-year-old male twins, separated between eight*
*days and six weeks. Minimal contact before their visit to*
*Minneapolis. The only black twins in the study and the*
*youngest*

Both twins had stuttering problems which improved over
   the years
Both had bedwetting problems which cleared up in one twin
   when he was seven and at twelve in the other; the second
   twin also soiled his pants until he was six
Both are extremely shy
Both started to speak later than other children, but the
   problems cleared up without treatment

Twin B had also been diagnosed as hyperactive and his parents thought that he had minimal brain damage. This was confirmed by the neurologist during the twins' week at Minneapolis. This twin had also shown several antisocial traits: he was known as a liar and thief, and had run away from home several times. On the other hand, twin A, who also had minimal brain damage, was emotionally stable.

Here again, although the details are slight, both twins seem to share many aspects of their psychiatric history – and with one rather more disturbed than the other.

## Patterns in the twin lives

This completes the account of the twins themselves. In some cases we know quite a lot about them; in others there is much more to be revealed. Already we can note with some confidence *three* important things which arise from the Minnesota study:

The similarities themselves fall into three categories. First, the purely anecdotal coincidences – the similarities in names, dresses, shirts, jewellery, cars and holiday spots, the same favourite authors and colours. Second, the psychological or behavioural similarities that have been measured and analysed – the same dreams, the same jobs, the same fears. Third, the psychiatric similarities – the same symptoms, of putting on weight, getting depressed, drinking a lot, being violent and so forth.

The first type of coincidence, the anedotal similarities in manners, habits and gestures, accounts for most of the public interest in Bouchard's study but is the most difficult to handle scientifically. If the medical similarities between twins mean that many psychiatric and even psychological similarities must follow, some extraordinary coincidences are to be expected. But what can explain the same names given to children, getting married on the same day, buying the same diary and filling it in in the same way? Are these also due to the identical genetic make-up which MZ twins share? Or is there really some kind of 'uncanny psychic bond' between twins, as Ethel Gaherty, for one, seems to think? Scientists have yet to evolve a proper technique for analysing coincidences and the rest of this book tries to do just that – to work out a way of assessing just how 'uncanny' the bond really is.

Although many twins share some psychiatric symptoms, in many cases one twin shows them in a far more severe form than the other. Twins often compete for nourishment in the womb, and this rivalry is more intense for MZ twins than DZs. It can result in quite sizeable differences between twins on their entry into the world. Are the differences between twins evidence of a permanent impairment at birth? Or can the environment markedly exaggerate a certain genetic predisposition so as to make a manageable disability in one twin something for which the other needs medication or hospital treatment?

Thirdly, there are some genuine, large differences between our twins. The significance of this is every bit as important as the significance of the coincidences. Some similarities may not be quite what they seem. For instance, both Gladys and Goldie started to use amphetamines for weight control 'between the ages of thirty and forty'. Now, this could mean that they started using the pills nearly ten years apart. How much of a coincidence is that? Is it a coincidence at all?

Oskar's and Jack's blue shirts were of very different blues and Oskar wore his over a roll-neck sweater, making his appearance quite different. Barbara and Daphne both fell down stairs when they were fifteen, but their accidents could have been months apart. Dorothy and Bridget both filled in their diaries for the same year – yet the entries are quite different. Terry and Margaret married on the same day – yet other twins who have met them say they do not even look alike.

Coincidence, like beauty, may be as much in the eye of the beholder as anywhere else.

*Part Two*
# The 'Natural Bond'

# 3  *Twins as Double Trouble*

Nowadays, our imaginations are caught chiefly by identical twins and the possibilities for confusion and amusement which they open up. But this does not always appear to have been the case. Although some twins must always have looked alike, it was not until 1865 that Sir Francis Galton predicated the difference between MZ and DZ twins and began to look for the consequences of this distinction. Until that time the main concern with twins (there were exceptions, notably in Shakespeare) seemed to be their poor chances of survival compared with children born singly. Twins, before the days of modern medicine, posed a threat to their mother's health, and their own survival was more risky, too. Occasionally, it was acknowledged that twins offered twice the pleasure for, as it were, only half the work. But only occasionally. Even now, with all the help of modern medicine, twins still pose problems for parents and doctors which babies born singly do not.

## Twin myths – a wet nurse from the king

Twins have often featured in religious history or as gods in mythology. After all, the specific purpose of religious stories and myths is to concern themselves with the doings of remarkable people. What is more interesting is that these early accounts usually stress not so much the similarities between twins as their differences.

In Scandinavian legend, for example, the twin gods Balder and Hoder fought over a goddess, the blind Hoder eventually slaying Balder. The Bible stresses the difference between Cain and Abel, each of whom had a twin sister, and who of course are chiefly known for their quarrel. Jacob and Esau, twin sons of Isaac and Rebecca, struggled within the womb over who would be first to emerge into the world (and Esau was hairy while Jacob was smooth). Two other biblical twins, Pharez and Zarah, grandsons of Jacob, also fought in the womb over who should emerge first.

Several tribes of American Indians had gods who were twins, including the Crow, the Navaho, the Kiowa, the Zuñi and the Shoshoni. The Pueblos had their own Adam and Eve myth about the creation, in which the original pair were supposed to have had five pairs of twins, a legend repeated in simplified form among other Indian tribes. One of the most widespread myths among these tribes recalls the biblical legends which emphasize the conflict between the first-born and the second-born twin. In many Indian stories the first-born is generally the hero, whereas the second-born is the villain and often has recourse to magic or sorcery. Among some Zulu tribes only the first-born used to be named, the other twin being left nameless, a practice reminiscent of that among certain Californian Indians, where the second-born would be smothered at birth. This may show an understanding of the problems of rearing two souls at the same time among peoples for whom even normal existence was a continuing struggle. The West African Ashanti tribe confined their butchery to very special twins. They only killed twins born into the royal family, resolving the problem of succession by killing off the younger twin at birth.

Many tribes, however, took an even more robust attitude to the birth of twins and, quite simply, killed both. The Zuñi and the Navaho likened twins to the litters of lower animals, a kind of evolutionary excuse. This practice occurs among all sorts of tribes – from Australian aborigines and South American Indians to Eskimos and the Ainu of Japan. In extreme cases in Africa even the mother of twins was put to death or at least made to undergo purification rites. (So pervasive was this practice that it was a common insult to say to a woman: 'May you become the mother of twins!')

A mother was vilified not just because she had brought upon the family a double economic burden but because it was felt that the birth of twins meant she had had intercourse with two men at about the same time. In monogamous societies this was greatly frowned upon. Accusations of sexual impropriety were also attached to brother–sister twin pairs, for many societies believed that they had had intercourse with each other in the womb. This at least was one reason advanced for the killing of boy–girl twin pairs at birth in Bali. Elsewhere the same beliefs about sex in the womb were current but the consequences less drastic: among the Bantu in Africa boy–girl twins were thought to have been 'married in the womb' and in the Philippines and in Japan boy–girl twins were actually required to marry when they reached maturity.

The tribes where twins are favourably regarded are fewer, though some African tribes, especially the Yoruba in Nigeria (where the twinning rate is exceptionally high), carve little figures, six or seven inches high, out of wood, to represent twins. The Yoruba believe twins to be good-luck omens, the manifestation of spirits who have entered a mother's womb in order to be born as twins. However, these figures, which are fashioned after the twins are born, do not really come into their own unless one or both of the twins die. Then the figures are 'cared for' in place of the dead twin(s) and carried about, either by the surviving twin or by the parents. The figures are fed, washed and dressed just like the real person they have replaced. This custom seems to reflect the greater risk to twins – they are so much more likely to die than other children that a charm or doll must be prepared in advance, ready to take their place. So the custom hardly indicates a totally positive attitude towards twinship.

The only groups where twins are an unmitigated happy event seem to be: the Dahomeans of West Africa, who have an elaborate ceremony when twins are born; the Ewe (also from West Africa), where the mother wears a special badge to announce to others that she has had twins; the Berber Moors of North Africa, where mothers of twins are addressed with the rank of nobility; and the Benin of Nigeria, where the king used to provide a wet nurse for the mother of twins, again as part of an elaborate celebration after their arrival. The Baronga of south-eastern Africa and some Canadian Indians also regard twins more favourably, owing to their association with fertility.

In general, however, the picture is consistent: twins were seen more often as double headache than a double blessing. No doubt these legends and myths unflattering to twins come from circumstances in which there was such little medical care that twins were seen as a health hazard, not just to themselves but to their mother; and if she died, they became an economic burden on the rest of the tribe.

## The children of the sky

Delagoa Bay, in south-eastern Africa, is a peaceful, remote, sunny stretch of coastline. Too sunny. The Baronga, the Bantu tribe whose traditional homeland it is, often have a problem getting enough fresh water. And that is where the twins come in.

In the spring, when the rains are due (and especially when they are overdue), the women of the tribe perform a special ceremony to bring down the rain. They replace their normal clothes with petticoats made of grass and leaves and sing what the anthropologists call 'ribald' songs; also they appeal for help to a family in which twins have been born. To begin with the women visit the house of a woman who has given birth to twins: there they drench her with (scarce) water which they carry in pitchers. The mother of twins is known in the village of the Baronga as *Tilo* – the sky – and the twins themselves are called 'The Children of the Sky'.

Next the women visit the graves of twins and drench those with water. It is their belief that the grave of a twin must always be moist, and for that reason twins are usually buried near the lake. If all efforts to induce the rain fail the women move the remains of a buried twin that is farthest from water closer in. But this is the last step: it is only ever done as a last resort.

The belief that twins possess some magical power over the weather is widespread and by no means confined to Africa. The Tsimshian Indians of British Columbia in Canada call the wind the 'breath of twins', and several other Canadian Indians believe that twins control the weather.

Usually the beliefs about twins and the weather are associated with ideas about water; the Kwaikiutl and the Nootka (also Canadian Indians) think that twins have a special relationship with, or can be transformed into, the salmon. In some instances this associates twins with plenty; in others it means that twins are not allowed near the water for fear they will be changed into the fish.

Curiously, no early record reveals an interest in twins as *identical* human beings. Identical twins must have been a source of confusion for others and of amusement for themselves. Yet nowhere do we find mention of this either in primitive mythology or early medicine.

One or two playwrights and authors have used the confusions that may stem from identical twins as entertainment, notably William Shakespeare. Possibly because he had twin children himself (a boy and a girl), he wrote about twins in no less than nine of his plays. In two, the *Comedy of Errors* and *Twelfth Night*, they are at the centre of the plot.

The scientific significance of twins first became apparent just over a hundred years ago.

## Galton's guess

It was a cousin of the great biologist Charles Darwin who first spotted the potential of twin studies. Sir Francis Galton began publishing his studies of twins in 1875. A connection had already been observed between a mother's age and the chances of her having twins, and several cases of Siamese twins were known in detail. Galton therefore made a guess that 'look-alike' twins came from one egg and contained the same genetic make-up, whereas 'look-unalike' twins, as they were then called, were from different eggs.

Galton was absolutely right in his guess, though the study of twins was hampered by the inability to test properly whether twins were monozygotic or dizygotic; and, since in many cases fraternal twins are more alike at birth than identicals, this naturally confused early students of twin similarities. It was quite common for MZ twins to be classed as fraternals at birth and then proceed to grow more alike as they got older, whereas the DZ twins, classified as identicals at birth, grew steadily less alike.

Over the past 100 years, our scientific understanding of

twins has gradually increased. That knowledge has developed in several directions; the main developments are here presented chronologically:

*1877*   The idea is first mooted, by M. Tchouriloff in France, that maternal height and twinning rate are linked. Since height was known then to be an inherited characteristic, this was the first attempt to link twinning and heredity

*1898*   Confirmation that twinning rate is highest in older mothers, especially those aged over thirty-six

*1905*   E. L. Thorndike, one of the most famous of the early psychologists, begins to compare the performance of MZ and DZ twins on IQ tests. This was the first flowering of that hardy perennial: 'IQ – how much is inherited?'

*1922*   The first study of identical twins separated at birth and reared apart

*1932*   The Maxim Gorky Institute in Moscow carried out a lot of work on twins, designed to explore the relative influences of heredity and environment. The programme of research was stopped in 1936 when it was held to be in conflict with Communist ideology. Soviet psychology was severely curtailed until the early 1960s, when it was 'rehabilitated'

*1933*   Early studies of criminality in twins

*1934*   Birth of the Dionne quintuplets in Canada. They are, so far as we know, the first quins to survive and the only all-identical set

*1934*   Early studies of schizophrenia in twins

*1937*   A British doctor, Sir Lionel Penrose, publishes studies of a pair of MZ twins, only one of whom had congenital syphilis – early proof that circumstances in the womb can be different even for identicals

*1937*   One of the first major studies of twin intelligence, in Chicago, concludes that IQ is largely inherited

*1953* Medical Research Council (London) publishes a special report on mental illness in twins which concludes that there is a major hereditary influence

*1962* British study of identical twins reared apart. Largest study of its kind reports that intelligence and mental illness have a large hereditary component

*1974* Founding of the International Society for the Study of Twins

*1978* Studies of Sir Cyril Burt, one of the world's most prolific psychologists, shown to be fraudulent. These studies 'using' twins were the basis for several social policies in Britain and helped Sir Cyril gain his knighthood

Twin studies have repeatedly been dogged by controversy. At the outset they were subjected to much the same vilification as Darwin's theories about evolution; in the 1930s their results proved unpopular with Soviet Communist ideologues; and most recently one of their chief protagonists, Sir Cyril Burt, has been shown to have invented at least some of his results concerning twins (mainly those which sought to show that intelligence was almost completely inherited).

Many scientists share the view that the debate over heredity, environment and their relative influence on people have been obscured, on the one hand by psychologists intent on exaggerating the effects of environment and, on the other, by geneticists exaggerating the influence of genes. Despite this, many of the basic facts and figures about twins are not at all controversial, after a hundred years of study. By and large, and despite medical advances, those figures bear out the ancient view that twins, though they may have twice the blessings, are rather more than double the trouble. Few twins are seriously handicapped, and growing up as a twin has its advantages. Nevertheless, the formal statistics do indicate that on average twins are born weaker, smaller and slightly less intelligent than non-twins. It is important to be clear about this because it may after all determine how twins are treated and affect the bond between them.

## Basic statistics on twins

There are an estimated 100 million twins in the world, and about a third of these are identical. The rate for identical twins is fairly stable around the globe – about 3.5 identical twins occur for every 1000 live births. The occurrence of fraternal twins is far more variable and depends on such things as the age of the mother, diet and race. Fraternal twins – but not identicals – run in families.

In many mammals litters are fairly large (think of dogs, rabbits, pigs). On the other hand twinning seems for some reason to be linked to the size of the animal; while mice, cats and small animals generally have multiple pregnancies, larger beasts – gorillas, horses, elephants, giraffes and cattle – rarely have more than one offspring at a time, just like us.

One sometimes reads reports about 'the magic number' regarding twins or the 'rule of 87' (or 86 or 90, whatever the newspaper editor happens to alight on). This usually refers to the 'fact' that, very roughly speaking, twins are born once in every 86 (87, or 90) births. As a broad – very broad – generalization, this is true; but it conceals such a wide variation from country to country, age to age, race to race, as to render it almost meaningless.

In Ireland, for instance, women have tended to get married late (until recently the average age was twenty-five to twenty-six). Add to that a religious aversion to artificial birth control and it is not surprising that many Irish women go on having children well into their twin-prone late thirties. The twinning rate in Ireland has run as high as one in every 67 confinements.

But even this is not high compared with some Negro tribes (Negroes have a higher twinning rate than whites or the Asian ethnic groups, who have the lowest rate). In West Africa the overall twinning rate is roughly one twin pair in 40 pregnancies, and the Yoruba in Nigeria manage a rate of one twin birth per 22 pregnancies. (The rate in Nigeria first gave rise to the idea that diet may have something to do with twinning because the concentration of twins there is said to vary with the consumption of a particular type of yam.) Negroes in America, who have a vastly different diet from the West

Africans and include whites and other races among their ancestry, have a twinning rate of one pair in 73 births – much closer to the white rate.

Among whites the rate varies also. It is low in Spain (one in 110 births) and high in Finland (one in 65). The USA and Britain both have a fairly low rate of twinning, nearer the Spanish level.

There are also variations in the ratio of identicals to fraternals. American Negroes, for instance, produce 70 fraternals for every 30 identicals, whereas for American whites the ratio is 65 to 35. On the other hand, in the case of the Japanese, who have a low overall twinning rate of one in 160, *identicals* actually outnumber fraternals by two to one.

## The ego of the twin

Although twins are rare, they are common enough for all of us, at some time or another, to have met them. Like famous people you occasionally come across in the street or at the airport, twins are *just* common enough to be interesting to the rest of us. And being an identical twin opens up all sorts of issues, some of them serious, some merely fun. The most intriguing is that twinhood, especially identical twinhood, faces us with people who, though separate individuals *biologically, psychologically* are not. We have all wondered what it must be like to be someone else. In our depressed moments we may even have wanted to *be* someone else. Twins cannot be someone else but they do have an altered sense of self. There is someone else with the same set of genes: unlike the rest of us, they are not unique. A study of MZ and DZ twins in Rome found that whereas MZ twins nearly always describe themselves according to the things they share with their co-twin, DZ twins take their identity from the differences. The Rome results suggest that MZ twins may 'go looking' for similarities and may even unconsciously invent ones that are not really there.

A third reason why twins are fascinating stems from the biological fact that, because they were two embryos inside the mother, they are far less likely to be as healthy as singleton children. The paediatric details about twins can be encapsulated in just a few paragraphs.

Life, for both identical and fraternal twins, is more complicated than for singletons right from the word go. In the womb an embryo is surrounded by a sac known as a chorion. Sometimes twin embryos share the same sac, which can be dangerous. The two umbilical chords of the twins may become entangled, one twin may crowd and injure the other one. They may compete in the womb for nourishment or may even 'jockey' for position, one draining blood away from the other. In all these cases the twins may show the effects at birth: although they are 'identical', one at first looks quite different, bigger, healthier, more advanced than the other. Another accident that can happen is that one growing twin foetus 'absorbs' the other. Cases like this are discovered only much later when, as an adult, an individual has an operation, say for the removal of a growth, and the surgeon finds a foetus mummified inside the body. It should have been a twin – but lost the race very early on.

In the majority of cases twin labours are not severe. However, all but 4 per cent of singleton children emerge from the womb in the easiest and safest way, head first, but this is true of only half the twins. According to research by Dr Alan Guttmacher and Dr Schuyler Kohl in New York, in 37 per cent of twin births one baby comes out head first and the other is breech (buttocks first). In a much smaller proportion of cases both babies are breech, and one of the twins may even lie crosswise in the womb as birth approaches.

Twin pregnancies are shorter. The average gestation for twins is 22 days less than for singletons (258 days as against 280). And twins are much more likely to be underweight and premature. More than half the twins born in the United States (about 54 per cent) weigh less than 5½ lb at birth and are classified as premature.

Premature twins actually have a greater chance of survival than premature singletons, but premature twins are much more common. It varies of course – girls have a better chance of surviving and the first-born is also less likely to die; and if twins survive their first month, their chance of survival then becomes just as good as for singleton children. But twins are more likely to die at birth than singleton children (and this risk is even greater for triplets, quads and quins).

Nowadays the difference in birth chances between twins and singletons has been reduced; incubators, especially, have proved their worth. It is perhaps understandable therefore that we have lost the fear the ancients had about twins, of the harm they thought they could do and the exotic ideas about their sexual relations in the womb. In its place we hold an equally exotic one, that there is a bond between twins, based not on their shortcomings but on their similarities. It is a more positive attitude – but is it more accurate?

# 4 *Parallel Lives*

Until Michael Chisholm joined the merchant navy as a cabin boy in 1955, he had been inseparable from his identical twin brother, Alex. They had lived with their mother and stepfather at Coatbridge, Lanarkshire in Scotland, for sixteen years and had dressed alike and played the same games. According to their parents and friends, they had never quarrelled.

When Michael sailed from the Clyde for Egypt, Alex, who was the older twin by a matter of minutes, went to see him off. That was on 28 December. Four days later Alex, who was then an apprentice, was celebrating New Year's Day with friends near Glasgow when he suddenly complained of tiredness. He was a strapping, healthy young man of seventeen but that night had to lie down on a couch. He died soon after of a heart attack.

Michael was in the Bay of Biscay when he heard the news; he immediately cabled home expressing his sorrow. But that night, Michael, just as well built as his twin, died in his sleep. It was less than forty-eight hours after Alex had died.

A bond between the twins? Stories of this kind are by no means uncommon. Tuula and Marietta Jaavaara, from Borgaa in Finland, had always done everything together. If they were separated for any length of time as children, they would fall asleep until they were reunited – for days on end if need be. Once, when they were separated, one of the girls staying with their grandmother, they both developed abscesses under their chins: both had to be operated on. Then, in December 1970, when the girls were twenty-three and apparently healthy, Tuula collapsed at home. Soon after, Marietta also collapsed, and their parents decided to take them to hospital.

Doctors tried to revive them by heart massage – there was no apparent reason for the collapse – but within an hour of their arrival at the hospital Tuula was dead. Marietta died ten minutes later.

Dorothy Collins lived in Brighton, England; her twin sister Marjorie lived in South Africa. In April 1961 Dorothy died from an accidental overdose of sleeping pills only a few hours *before* a cable arrived at her house with the news that Marjorie had just died. Jim and Arthur Mowforth, also inseparable as boys and with almost identical RAF careers, died on the same day in April 1975, aged sixty-six. Jim died from a heart attack in Bristol, and Arthur from the same cause in Windsor, ninety-three miles away.

The fact that *some* twins die on the same day is not surprising since on *average* identical twins have very similar life spans. If one dies in infancy the other tends to do so as well: if one dies in middle age, so does the second. This is a fact of medical life and, although the mechanism whereby it happens is not understood, there is nothing really 'uncanny' about it.

There are several reports of one twin influencing the feelings of the other. For instance, Mr Cecil Couper of Winnipeg, Manitoba in Canada, was not expected to live after a heart attack in June 1949. He had good medical attention but grew worse day by day. As he grew weaker, sinking into unconsciousness, he began to call out for his twin brother, Leslie, whom he had not seen for thirty-two years – since 1917. His sister, Mrs Nora Reynolds, heard of this and remembered the affinity between the two men when they were boys in England. She called Leslie and, in four days, he was at the bedside of the brother he had last seen as a soldier of twenty-six. Observers recorded that, from the moment Leslie arrived, Cecil began to feel better. A month later he was walking around the ward of Deer Lodge Hospital, on the road to recovery.

John and James Cramp attracted attention in 1955, when they were aged three. Johnny, even though out of sight of his brother, would laugh when Jimmy was tickled. Put back to back and asked to draw whatever came into their head, these twins usually drew the same picture – and announced, at the same time, 'I've finished.' Hidden from each other and given the same box of chocolates, they chose the same sweet.

Keith and Kenneth Main were very similar, too. On 13 November 1958, Keith underwent an exploratory operation at Newcastle-upon-Tyne General Hospital in England because he had a hole in his heart. At that time, his twin brother Kenneth ran sobbing to their mother in near-by Gateshead, complaining of pains in his chest. The time was 12.15 p.m., the exact time – the Mains found out later – that Keith was on the operating table. And a few days after that, when Keith had his stitches removed, Kenneth cried out again. So severe were his pains that the parents took him to see a doctor. The doctor could find nothing wrong.

Not only did these twins appear to share each other's pain, they caught the same illnesses at the same time. 'We have three other children,' says Mrs Main, 'but it's always the twins who go down together with chickenpox or whatever.'

Doreen and Ella Colley had different things in common. In 1959, when they were eighteen, both appeared on remand at Nottingham charged with stealing a jumper from a store. It emerged at the hearing that both twins had almost identical criminal records. Both had two periods of probation, both had been fined and both had been to an approved school. Both had four convictions for theft.

Brian Blackett was walking alone in the woods in Epping Forest, to the north-east of London, on Thursday, 13 October 1960. Suddenly he was frightened. 'My legs began to shake,' he was to say later. At that moment, twenty miles away, his twin brother Lennie was being sent to prison for three months on a housebreaking charge. Brian, the younger twin by thirty minutes, said, 'I knew something was wrong. I didn't know that Lennie was coming up today. I felt very weak. I wondered what was happening to my brother.'

The boy's father, Mr Leonard Blackett, appealed to the chairman of the court not to separate the fifteen-year-old twins. He told the court that Brian was suffering at home because he was separated from his brother, who was on remand. 'I have known one of the boys fall down the stairs and the other one to weep for him,' he said. The sentence, however, stood.

Barbara Morgan, a nineteen-year-old student, had labour pains for six hours on the evening of 26 March 1969. She was given sedatives by her doctor, but it was her twin sister Gillian who was having the baby. Gillian was moved into the delivery unit of a Manchester hospital while Barbara was visiting her. And, when Gillian was given pain-killing injections, Barbara went to sleep too. From childhood the sisters had swopped pains during illnesses and, the year before, Barbara had developed morning sickness and later backache. She had written to her twin, then in Germany with her husband, asking 'What's wrong with you?' That was how Gillian found out she was pregnant.

Mrs Martha Burke felt so strongly for her twin that she even sued the airline when her sister was killed in 1977, in a 747 collision in the Canary Islands. Mrs Burke was 6000 miles away at the time of the crash, in Fremont, California, but she claimed that she felt a 'terrible burning sensation' in her chest and stomach and knew that something terrible had happened to her non-identical twin sister, Mrs Margaret Fox.

Andrew Huxtable, honorary secretary of the National Union of Track Statisticians, has studied the athletic performance of identical twins. He has found a number in Britain, Finland, Belgium, Iceland and the USA who have recorded remarkably similar performances in various events. The most striking case was that of David and Robert Holt, whose best times for the six miles were almost identical – 27 minutes 43.6 seconds, and 27 minutes 43.0 seconds, respectively. Over six miles, 0.6 of a second represents 11 feet 4 inches – that is how close the Holts were.

Finally, twin similarities of a totally different kind. Greta and Freda Chaplin, who live in York, England, seemed to be 'trapped in one mind'. They had both fallen for a lorry driver who lived next door to them. Their obsession, which continued over many months, finally ended up in the courts when the lorry driver wanted himself rid of their attentions.

During the court case in November 1980 the bizarre – and far from funny – pattern of the Chaplin twins was disclosed. At thirty-seven they were unmarried, had no home of their own, and no jobs. Every morning they got up at the social services hostel, where they had one of the few double rooms.

They would get ready together in the bathroom; if given different cakes of soap they would cut them in half and swap with each other. They had the same breakfast – egg and tomato – and both would have their hands on the frying pan *together*. They dressed alike, down to stockings, underwear, headscarves and buttons, always shopping at a store with many identical clothes. They spent their days in a hospital, pursuing occupational therapy. There had been attempts to separate them, but they had simply defied them and got back together again. (They would scream if separated as young girls.)

Spending all their time together attracted ridicule, for example, from children they met in the street, so they did not go out much and preferred the winter when the dark nights meant they were not so noticeable. A dismal, strange life, but they clung to it.

But enough for the moment of the anecdotal evidence for the twin bond. Note though that in almost every case a more down-to-earth explanation for this bond is possible than telepathy or ESP or clairvoyance. It is not unknown for husbands to have labour pains when their wives go into hospital; you would expect similar athletic performances from people who are very similar physically; and is it really so surprising that one twin influences the recovery of another? Recovery from illness is known to be affected by many emotional factors. In the case of the twins who felt pain when their co-twin was operated on or the boy who laughed when his twin was tickled, we do not know the exact circumstances of the episodes. One twin may sometimes know something about what is going to happen; there may also be unwitting signs between twins that offer clues as to what each one is about to do. Even circus animals can 'read' the unconscious mannerisms of their trainer which tell the animal when to do a particular trick: a famous case was a horse that could 'count' by tapping his hoof on the ground. The trainer would call out a number; when the horse got to that number, a slight movement by the trainer, of which he was unaware, told the horse to stop.

In the case of the Chaplin twins, it must be said that they were well below the normal level of intelligence and showed not so much an 'uncanny bond' as a dependency that is seen in many patients in hospitals for the subnormal. The difference was that they were dependent on each other. (Dr Hugh Gurling, of London's Institute of Psychiatry, believes that the Chaplin twins are suffering from Erotomania – De Clerambault's Syndrome – which is a variant of schizophrenia. He and his colleagues have another, similar, pair.)

The case of Mrs Martha Burke is perhaps the most puzzling of the anecdotes, as it is the least amenable to explanation. On the other hand, with all due respect to Mrs Burke, we have only her word for her reaction on the day of the air crash and, for a scientist, that can never be enough.

The word twin derives from the ancient German word *twina* or *twine*, meaning 'two together'. The modern German word is *zwillinge*, from the same base as *zwei* – two. This book is, however, concerned with identical twins, not just two children born together at the same time.

Identical twins, as we have seen, are the result of one of the mother's eggs being fertilized by one of the father's sperm, and then splitting in two to form two genetically identical individuals. This is how identical twins have exactly the same genetic make-up. Non-identical twins are formed from two eggs, each separately fertilized by two different sperm, and they are no closer genetically than ordinary brothers or sisters.

Life in the womb for identical twins is unusual. They commonly share the same placenta and amniotic sac; in other words, they have the same prenatal environment, but this also means they have to compete for food and oxygen from the mother. And the rivalry between MZs for this food and oxgen is intense – much greater than that between DZs. This is one reason why MZs may be *less* identical at birth, in their appearance at least, than DZs.

In discussing the 'bond' between MZs, we shall see that parents, teachers and other people treat them in a different way to DZs. Before that, however, here are a handful of extraordinary stories that remind us how spectacularly independent of each other non-identicals may be. It is a useful benchmark when considering the lives of MZs.

## The black and white twins

In 1977 a German girl, Grete Bardaum, made love to a German businessman and then, later the same day, to an American soldier. Nine months later, in early 1978, she gave birth to 'twins'. I use the word in quotes because she had conceived a child from both men on the same day, and the boys were born together. How could she be so sure? Because one of the twins was white while the other, like his father the American soldier, was black.

In that same year, 1978, it was reported in the *Lancet* that a Swedish woman had given birth to a healthy child after doctors *deliberately* killed its twin in the womb because it was suffering from an *inherited* disease. By taking cell samples from amniotic liquid surrounding each baby, the doctors discovered that one was suffering from Hurler's syndrome or gargoylism, an abnormality which causes deformation of the face. Not only was the second twin free of the disease but, doctors judged, it had a very good chance of survival and being normal. Reluctant to lose both babies, the mother asked if one could be killed and the other left to develop alongside the dead foetus.

The operation had never succeeded before but the Swedish Health and Social Bureau gave the go-ahead. Dr Anders Aberg of the University Hospital in Lund punctured the heart of the diseased baby with a needle. The twins were then monitored regularly by ultrasound; and the normal one happily continued to grow while the other shrank a little in size.

Also in 1978 a California woman gave birth to twins – two boys – though the man whom she said was the father denied it. During the following court case a new test of paternity was tried, one which used many different antigens in the blood. This showed conclusively, according to Dr Paul Terasaki, of the University of California at Los Angeles, that the man who contested paternity was in fact the father of one of the boys *but not the other*. It turned out that though the boys were born on the same day they could have been conceived up to a month apart.

Finally, the case of Mrs Antonio Legaz, from Cartagena in southern Spain. On 3 April 1977 she gave birth to a daughter,

Maria Augustina. Then on 6 May, only a month later, she gave birth again – this time to Maria José. Now here is an interesting philosophical and biological question. Mrs Legaz, it turned out, had two sets of reproductive organs, including two wombs. No one ever knew whether the children had been conceived simultaneously or separately. But the babies had spent at least eight months of their respective pregnancies together inside their mother. Does that make them twins or merely a special kind of sisters?

## The birth of the bond?

The arrival of twins in a family has a surprising effect on the parents. According to a Canadian study of forty-six sets of MZ twins, published in 1978, parents speak much less to twins than they do to other children. They also show less affection – for example, they hug and kiss twins less than they hug or kiss their other children. On the other side of the coin they also seem to be more lenient in that they punish twins less. In these results one can perhaps see the beginnings of a bond between the twins *being created* by the way the parents react. Parents may be more exhausted when raising twins (there is a great deal of evidence from Mothers of Twins Clubs for this); consequently, they may speak to them less, and have less energy or inclination either to show them affection or to punish them.

An interesting study of twin families was recently completed in Jerusalem by Esther Goshen-Gottstein. She looked in detail at four families of twins, six families of triplets and four families of quadruplets, in all cases from the ages of five months until six years. She found, first, that several of the mothers were ambivalent about having more than one child at a time. Some, even today, felt a superstitious reaction, imagining that it was a religious punishment. One woman said specifically that her twins were due to the fact that, secretly, she had not wanted any children at all. Immediately after the birth, most of the mothers had very negative feelings about their children. In some cases this persisted, in others it disappeared only to recur later, when the twins or triplets were several months old and new forms of behaviour began to be a problem. Several of the mothers showed their ambivalence to their twins by going out a lot – and staying out, leaving the twins alone. In other cases the children were physically confined to a playpen or a room – and then the mother would avoid that room or area. One of the mothers of quads refused to be seen in public with them: she always insisted that her husband go off with two of the babies and she took the other two in a different direction.

The point about all this is that the mother's behaviour could well drive the twins in upon themselves and force them to seek from each other the stimulation and comfort not forthcoming from their mother. Certainly, Goshen-Gottstein reports that twins tend to clamour for their mother's attention far more than other children – as if they are being starved of it.

After this ambivalent period, Goshen-Gottstein found that many mothers began to treat their children as a single unit. This again seemed to be rooted in the very practical, everyday fact that two children (or three or four) are naturally more of a handful than one. But some of the examples she gives underline how behaviour by the mother of twins can highlight the similarity between them – and may therefore *create* the appearance of more similarity.

The two, three or four children would be made to use the same handkerchief even though only one child would have a cold (not for long, no doubt). All would be put to bed though only one was tired. All would be fed solids even though one child had not yet cut enough teeth. One would wet his or her clothes but all would be changed – very often to ensure that they were always dressed the same.

Whatever their reception, however, by the time twins (or triplets) cease to be babies, most mothers have come to enjoy twindom. Several studies have shown that mothers rejoice when their twins cannot be told apart by other people; in the Israeli study several mothers claimed that their twins had started talking and walking at the same time and in the same way, even though trained scientists, admitted to their homes to observe objectively, found this was simply not true. And mothers talked about the twins or triplets or quads as a single unit – using just one name, or 'he' or 'she' for a set of boys or girls. When a mother's triplets were made up of a pair of identical twins and a third, non-identical, she would refer to the identical twins as a single unit, stress their similarity and highlight the differences with the third baby. The overwhelming pressure to similarity can be seen in twin boys, aged three, who were judged by most of the observers to be very different from each other – yet the boys were so confused about their identity that they would each react to either name.

Goshen-Gottstein feels that the mothers' treatment of twins and triplets shows some biological factor at work which ensures that a mother can respond optimally only to one offspring at a time. She feels that such a mechanism would be valuable in the evolutionary sense, since it would help ensure the survival of the stronger baby wherever a multiple birth occurred. Such a mechanism might explain the pressure to treat twins as a unit. Goshen-Gottstein also reports differences in the mothers' treatment of their infants. During the ambivalent phase, for instance, there was a tendency for some mothers to see only good in one twin and only bad in the other. She suggests that this may be another way some mothers simplify their relationship with their twins – by concentrating for a while on one twin and making him or her the favourite.

The overall message is clear: biologically, humans have developed to the point where a single offspring offers the best possibilities for survival and, where twins or triplets occur, there is a tendency for a mother to treat the multiple birth as a single unit. It then follows that the similarities of the offspring are stressed in this process. This may well be the birth of the bond.

Certainly, when twins sleep with their thumbs in each other's mouth, or take it in turns to cry to attract their mother's attention, they are *behaving* as a unit. Once set up, this pattern may be hard to break; many parents have tried to get their twins to sleep separately, only to discover, the next morning, that one has crept into the other's bed.

In these early years parents may actually impede the development of a sense of individuality in twins by regarding them as a 'unit'. The twins, of necessity, tend to learn from one another, rather than from a parent or an older brother or sister. The evidence shows that twins speak less than single children and use a less advanced vocabulary, age for age. This finding has been confirmed by several IQ studies, notably in Birmingham, England, where the records of 2000 twins and 4800 single children were compared. The average IQ of the twins was 95.7 compared with 101.1 for the single children.

Twins have been known to develop a language of their own. Perhaps because their parents are too exhausted to talk much to them, twins do not learn to speak properly and fall back on each other, using their own mixture of new words and pidgin. This certainly seems to have happened in one of the best-documented recent cases. Grace and Virginia Kennedy were born in 1971; by the time they started school in California five years later they were listed as 'mentally retarded but trainable'. At first, it seemed that they could not talk; then they spoke gibberish. Only later was it recognized that they had developed a language of their own. In part it was a combination of English and German (their parents were English-speaking, their grandparents spoke German). For instance they said *milsh* for milk. But often vocabulary and grammar were all their own. When they had finished playing with a toy, for instance, they would say, *'No, no ingiddin.'* Although many twins have their own language to begin with, they grow out of it by the time they are three, but the Kennedy twins were still using theirs when they were six. Tests at school show that they are very bright: their special language is their only problem.

Another similar case in England was successfully treated by placing the twins, Alice and Beth, in different classes in school. At first they visited each other continually but gradually this lessened. During the second term they each had special language tuition for half an hour a day – separately of course. In eight months Alice's IQ score rose from 69 – very subnormal – to 105, just above average. Beth did rather less well, rising from 69 to 86, but she was still able to progress up the school.

## 'Twin' names – and dressed alike

Whatever the problems of twins in the early days, the special enjoyment parents find in their twins may be a factor in the twin bond. Take twin names as an instance. In one study of 340 pairs of twins, Dr Amram Scheinfeld found that four pairs in ten had what he called 'twin-type' names – with similar sounds or meanings: Ronald and Donald; Steve and Stanley; Sue Ann and Jo Ann; Nancy and Nora; Jack and Jill. This twin-type naming was more common with girls – over half of them were so named, but only a third of the boys. Some scientists think this can be confusing for the twins. Dr Irene Lezine, a French psychologist, found that, age for age, the proportion of child twins who did not know their names was 50 per cent *higher* than for singletons.

If this confusion has important lasting consequences (and it is by no means certain that it does), it is not helped by dressing twins alike. The picture here is much the same as with names. In Dr Scheinfeld's survey four out of five pairs of same-sex twins were dressed alike (again more frequently with girls). Of course, it may be cheaper to dress twins alike – and may also avoid the risks of favouritism. Even so, twins dressed alike are bound to attract more attention than those who are not, and are probably more likely to be treated as a single unit. So dress may be even more effective than names, in the long run, in keeping the bond between twins alive.

We can expect certain similarities to occur – and recur – as twins grow, simply because their genetic make-up affects several aspects of their lives. For a start, it affects their looks. Hair can be dyed, waved or cut, but identical twins have much the same stature (even allowing for diet), the same eye colouring, the same or very similar features. Similarly with illness. The case of a pair of identical twin doctors underlines the fact that MZ twins not only have identical eye colouring but the same hormone balance, the same blood make-up, heart rate, brainwave pattern, the same digestive tract and much else:

Both had spina bifida, present from birth
Both had an abscess near their tonsils, operated on when
they were in their early twenties

Both had arthritis in their little fingers
Both had the same eyelid inflammation
Both had the same degree of myopia – their spectacles were
   interchangeable
Both were sensitive to heat and preferred cold weather
Both had the same pattern of hair loss
Both had lost the same teeth (but on reverse sides)
Both had developed a heart problem in the same year
Both had bleeding duodenal ulcers; one had a gastrectomy
   (removal of part of the stomach) three years before the
   other
Both died within seven weeks of each other from the same
   illness, arteriosclerotic disease

More than this, the social consequences of their diseases made
their lives similar. Both had to retire at about the same time
because of their ulcers. Neither married his girlfriend since
they were both so uncertain about their health that they chose
not to risk marriage. (And who can tell whether this decision
was made earlier by the fact that each knew he was not alone,
that they had each other?)

   There *were* differences between the two: they went to different medical colleges; one went into the army for a while
but the other did not. But they are a good example of the
way the constitution of twins can affect their lives in similar
ways, via illness and medical defects.

   Then there is IQ, which continues to be a matter of enormous controversy. For the present I shall assume that IQ
tests really do measure intelligence, whether inherited or not.
What matters here is the effect that the very similar IQ which
twins possess has on any natural bond between them.

   The point is made simply in the figure opposite, which
shows the degree of closeness between twins and others so
far as IQ is concerned.

   The fact that MZ twins have very similar IQs surely affects
their lives – and the coincidences in them – in many ways.
We know for instance that many jobs have an optimum IQ
range: if you are below that range you will not be able to do
the job; if you are much above it you will very quickly find
it boring. It probably follows that MZ twins – whether they

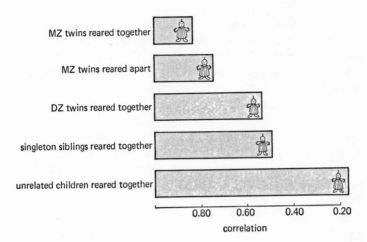

are reared together or apart – tend to gravitate towards jobs in roughly the same range. We know, secondly, that 'assortative mating' takes place: people tend to marry people with much the same IQ as themselves. So twins' spouses tend to be similar in this respect also. And of course educational levels also tend to be the same.

IQ tests are made up of sub-tests which reflect certain areas of interest, mathematical ability, spatial ability, creativity and so on. As the similar overall IQ scores of MZ twins seem to be made up from a similar pattern of performance in the sub-tests, it follows that they tend to be good at the same kind of things (good with their hands, say, but with no musical ability). This too must surely determine the kind of career they follow and the people – colleagues, friends, wives or husbands – they meet. It could even affect where they choose to take their holidays.

In terms of personality, much the same applies. Dr Raymond Cattell, a famous personality psychologist, studied twins aged from eleven to fifteen and found that, on such traits as being warm or cold, shy or outgoing, identicals were much closer than fraternals. In another study the same result was obtained for response to music, paintings and other cultural interests. And MZ twins are notorious with psychologists for the similarity of their likes and dislikes in food. The identical-twin doctors mentioned above, for instance, both hated peppermint, tomato soup and grapefruit. They had discovered this separately, but in many cases food preferences are apparent almost from birth.

As they grow into adult life some twins go their separate ways, but a number – perhaps a surprisingly large number – keep in close contact. MZ twins keep closer than DZs, women especially. Looks are also involved. Good-looking twins get treated throughout their lives as more of a novelty than plain twins, and their twinhood becomes more related to their success: they have more need to stay together. One Swedish study showed that MZs are about twice as likely to stay together as DZs.

By the time they are adults, many MZs are, in fact, closer to their twin than to anyone else. Professor Ernest Mowrer's study illustrates this best. He asked over 600 twins, 'Who in your family understands you best?' Sixty-one per cent of the MZs nominated their twin compared with only 39 per cent of same-sex DZs. Half of the MZs also said that they would miss their twin more than their mother if either died, whereas only 25 per cent of DZs put their twin first. And 30 per cent of female MZ twins said they would like to live *next door* to their twin after marriage. (I have found nine cases where MZ twins have married MZ twins. These cases are interesting scientifically because their children, although legally *cousins*, are genetically brothers and sisters. Where this has happened the children have felt themselves to be, and in fact been treated as, brothers and sisters.)

No doubt many twins are glad to go their own ways. Yet it is undeniable that several set up in business together (and even start Twin Clubs). And there have been notable twins in many walks of life: the scientists Robert and Wallace Brode, painters Raphael and Moses Soyers, journalists Abigail Van Buren and Ann Landers (who are actually Pauline Esther and Esther Pauline Friedman) US Air Force Generals Barney and Benjamin Giles, British Air Vice-Marshals Richard and David Acherly, and French Generals Theodore and Felix Brett.

## Coincident deaths

The trouble with the anecdotes at the start of this chapter is that they are just anecdotes. Are they really typical? Dr Franz Kallmann and his colleagues have studied the longevity of 1400 elderly twins (aged sixty and older). His results for the average time between deaths are as follows:

*MZ twins*
Male twins die 4 years and 2 months apart
Female twins die 9 years and 6 months apart

*DZ twins*
Male twins die 6 years and 3 months apart
Female twins die 10 years and 7 months apart

These figures make clear that the newspaper accounts are poor guides to the real situation. There are perhaps in all a dozen or so stories about twins dying on the same day and such sad coincidences, as Kallmann's research shows, are very rare. Further, the newspaper accounts do not suggest, as the figures show, that male DZ twins are more likely to die close to one another than female MZs. In this respect, the sex of the twins is a greater determining factor than whether they are MZ or DZ. Newspapers latch on to them because they are intriguing, but they are no more than that, any more than other coincidences such as the fact that the painter Raphael died on his birthday and that his birth and his death both took place on a Good Friday. Research shows that *most* MZ twins die a few years apart.

We now have an overall picture showing the chronology of the lives of twins and how those lives are similar. We can say that any consideration of an uncanny bond between twins has to take into account the medical and psychological similarities between them which, as we have seen, cause them in perfectly understandable ways to have similar life patterns.

## The differences between identical twins

Further to the natural links we have seen created between twins, their points of difference should not be overlooked.

Let us begin with IQ. We saw how similar it was for most MZ twins. Yet there have been at least two documented cases of MZ twins having IQs with about 20 points' difference. In one case, twin Gladys, who was raised in the Canadian mountains, had an IQ of 92; her sister Helen, raised in the Midwest of America, had an IQ of 116. So Gladys was marginally backward, whereas Helen was almost bright enough to go to university.

The Minnesota study commonly found fears and phobias in one twin but not the other. In two cases out of the sixteen pairs one twin had a fear of heights but the other did not. In another case only one twin had a fear of snakes. Bedwetting also seemed to vary. In two sets of twins both did it in childhood, but in a third pair only one twin did. In five sets of twins one was neurotic enough to require medication, whereas the other was apparently stable. Either Jack or Oskar had sleep problems – a tendency to keep falling asleep – whereas the other could be very violent and had even had psychiatric treatment for it. Neither showed any evidence of the other's problems.

Oskar and Jack are a good example. We saw earlier how similar they are in many respects. But in other ways they are very different. Such differences may not make headlines but, scientifically, they are equally important.

Let us end this chapter as we began it – with a fascinating story, and one which this time nicely illustrates the differences between MZ twins.

## The story of Chang and Eng

Chang and Eng were the most famous twins ever – the original Siamese twins. They died within two hours of each other. But, it can be argued, it was their differences that killed them.

They were born in Siam in May 1811. Their mother was the half-Chinese, half-Malayan wife of a Chinese fisherman and a thoroughly sensible woman. Rather than run away from her responsibilities she brought up the twins with a mixture of firmness and kindness. Although Rama II, the King of Siam, ordered them to be killed, she managed to keep them alive and to bring them up to be independent and as normal as possible. They were, for example, as good at swimming and diving as any of their normal friends.

They grew up to be short, wiry men, just over 5 feet in height, connected by a band of tissue 4½ inches long, running from the lower end of the chest to a common navel. But they were highly intelligent men, with numerous talents.

They became famous as a circus act after they had emigrated to America. By the time they were thirty they had earned enough to settle down in the rural community of Mount Airy in North Carolina. Within two years they had assumed the family name of Bunker and married.

Their wives were sisters, Sarah and Adelaide, the daughters of a local farmer in the Blue Ridge Mountains. These marriages, perhaps the most curious in history, resulted in twenty-two children – three boys and seven girls to Chang and Adelaide, seven boys and five girls to Eng and Sarah. Chang and Eng now have more than 1000 descendants, including a general in the US Air Force. For some years they all lived in the twins' original house; but when the families grew they built separate homes, Chang and Eng spending three days in one and the next three in the other.

Their latter years highlighted the differences between them – and brought a sadness. Chang's health began to fail, mainly because he started drinking. Eng become so worried that he tried to get separated, but no doctor would accept the operation.

The end was poignant. On Monday 12 January 1874, Chang took to his bed with bronchitis in his own house. On the Thursday, it was time to move, according to the arrangement, to Eng's house. Eng did not want Chang to go but the latter insisted. It was a bleak day and the move did not help Chang's health: he grew steadily worse. When Eng woke at dawn on Saturday 17 January, he could get no response from Chang. He called for help but was told by one of his sons, 'Uncle Chang is dead, father.' Eng replied, 'Then I am going, too.' He did, two hours later.

Before we look at the 'uncanny' coincidences, five main conclusions need to be borne in mind:

The way twins differ from the rest of us, and the way they are treated as a result, may help them to think of themselves as a special unit. This could be the basis of a bond.

Twins, being a greater strain for parents than single children, are brought up in a different way. This may so affect their development and psychology as to reinforce their idea of twindom, so much so that their twin becomes the most important person in their lives.

The simple medical facts of life – which MZ twins share – can have far-reaching effects, not just medical, but financial, occupational, social and educational, so that the life styles of twins, separated or not, are likely to be very similar.

Anecdotes about twins, especially those picked up by the popular press, may be true in a particular instance, but the impression they give is, quite simply, misleading. For instance, newspapers tend to publicize the fact that twins have a remarkable tendency to die on the same day, but this is not borne out by statistics.

As well as the similarities, the differences between twins are there too, if you look for them. There were many similarities in the lives of Chang and Eng – there had to be since they were joined together at the chest. But only one became alcoholic and, in killing himself by drinking, killed his twin too.

*Part Three*
# The 'Uncanny Bond'

# 5   The Science of Coincidence

## The world is not a small place

'I believe that there are 15,747,724,136,275,002,577,605,653, 961,181,555,468,044,717,914,527,116,709,366,231,425,076, 185,631,031,296 protons in the universe and the same number of electrons.' That was how Sir Arthur Eddington, the distinguished mathematician, physicist and astronomer, started one of his scientific papers in 1939. As Keith Ellis points out in his book, *Number Power*, it is an exceptionally arresting first sentence. Eddington wasn't certain of the figure; he thought he might have made a mistake in the calculation. But if there was, he said, it must have been a simple error – so he wasn't far out.

Eddington, like any mathematician, was interested in finding the rhythms in nature that govern our lives, but I mention Eddington's figure now to remind us that the world is *not* a small place. It is quite large enough to hold a great many things. For instance, the earth contains:
4,200,000,000 people;
14,900,000,000,000,000 trees; and
1,597,100 different species of living things.

Think of it another way: in an average year we catch:
25,060,000,000,000 fish.
In a single *day* we smoke 7,344,000,000 cigarettes.
At any one time, man shares his planet with an estimated:
3,000,000,000,000,000,000,000,000, 000,000,000 living things.
These are all large numbers and they do not even begin to cover the natural and man-made things in the world. With so much going on, therefore, coincidences really *should* occur from time to time.

This chapter is, in large part, about numbers, but the maths are not very complicated and the actual numbers should be fun: not abstract concepts but the numbers and proportions of everyday things. We need to look at numbers if we are to work out whether the coincidences between twins are anything more than chance – or whether they are so strange that they require an 'uncanny' or 'supernatural' explanation.

## Unlikely events do happen

In January 1963, the British actor Sean Connery (of James Bond fame) walked into the St Vincent Casino in Italy and backed the number 17 at roulette. It came up. He backed it again. It came up a second time. He backed it once more . . . and won for the third time in a row. In a few minutes he had netted himself £10,000. Now, no one suggests that he was psychic or that he was cheating. And he has never had that kind of luck since. Yet the chance of the same number coming up three times in a row at roulette is 1 in 50,652. (Looked at the other way, it is not an unusual event at all. Tens of thousands of people gamble at hundreds of casinos round the world each day – so this event, with a probability of only 1/50,652, happens *certainly* every few days, perhaps on most days, in one casino or another.)

Many examples of even rarer events actually happen without any suggestion of paranormal activity. One of my favourites is the Beatrice Choir Explosion.

Beatrice, Nebraska, is a small, farming town, little more than a village and very religious. On 1 March 1950, choir practice was set for 7.20 in the evening. The minister, however, was a little late that night. His wife and daughter were members of the choir, and he waited while his wife finished ironing their daughter's dress. A fourth member of the choir, a girl, was also late: she wanted to finish a geometry problem she had been set that day at school for homework. Two other choir members were late because their car would not start. And two others stayed home to hear the end of a particularly interesting radio programme. In fact, all fifteen members of the Beatrice choir were late that night: not one arrived before

7.30. Which was just as well, because at 7.25 an explosion destroyed the entire church.

This, as I said, is my favourite coincidence. There is no doubt that it happened, and the thought naturally arises that the choir's good fortune was, as one of them later put it, 'an act of God'. But . . .

Let us assume that any *one* chorister would be late about one choir practice in four. (Think of your own behaviour in getting to work on time or to any regular meeting you have to go to: one in four seems not too far from the truth.) In other words, there is a one in four chance that *one* member of the choir would be late on any given night. Now in this case there were actually ten reasons which delayed the fifteen members of the Beatrice choir (for instance, the dress in need of ironing was a single reason that delayed three members – the vicar, his wife and their daughter). We can say that the probability of everybody being late on the same night is: $(\frac{1}{4})^{10}$ or $\frac{1}{4} \times \frac{1}{4} \times \frac{1}{4} \times \frac{1}{4} \times \frac{1}{4} \times \frac{1}{4} \times \frac{1}{4} \times \frac{1}{4} \times \frac{1}{4} \times \frac{1}{4} = 1/1,048,576$. This is the same as saying that there is about one chance in a million that the ten reasons would all crop up on the same night.

So, if we leave aside the possibility that what happened that March night was an act of God, the important lesson of this episode is that events as rare as that, one in a million, *do* happen. The truth is, in fact, that rare events are happening all the time. If that sounds paradoxical, remember it is a large world.

Perhaps the most 'unlikely' gambling story that actually happened was when the 'even' number came up at a Monte Carlo casino *28 times* in succession. In these casinos the roulette wheel contains the numbers 1 to 36 *plus* a nought. So the chances that an 'even' number will come up at any particular time are 18/37, marginally less than one in two. Therefore 28 'even' numbers consecutively is a bit like tossing a coin and having it come down heads 28 times. This configuration, it has been calculated, would occur by chance once in every 268,435,456 times – *but it happened*. Mathematician Warren Weaver worked out, from the number of casinos and number of players each day, that, on average, this event should take place every 500 *years* at Monte Carlo. Rare events do happen.

Statisticians have developed ways of assessing 'chance' events. It often happens, for example, in a complicated world, that the link between a possible cause (say, smoking) and a possible effect (say, lung cancer) is by no means straightforward. (Not everybody who smokes dies from lung cancer.) The scientists work out the probability that the effect follows from the cause and see how it compares with chance. For instance, let us say that people who smoke are twice as likely to die from lung cancer as nonsmokers. Now statisticians know the chances of anybody dying from almost any disease at any age and they have agreed among themselves, for instance, that if the odds of death associated with a particular cause (smoking) are increased by up to 1 in 20 then no cause has been demonstrated – the effect is pure chance. If, on the other hand, the odds are increased by between 1 in 20 and 1 in 100 then something is happening, not by chance, but by the cause you have in mind. Between 1 in 100 and 1 in 1000 most scientists accept that this is not a chance occurrence, an effect is definitely being shown; and above 1 in 1000 whatever is happening is not happening by chance. So, if the death rate from lung cancer among smokers is 1000 times higher than for non-smokers, this is not chance: smoking causes cancer.

Here are the probabilities of some everyday and less than everyday risks so you can gauge for yourself:

| | |
|---|---|
| That you will be divorced | 1 in 3 |
| That it will rain today in New York | 1 in 3 |
|     in London | 1 in 3.2 |
| That your home was broken into | |
|     during the past year in the USA | 1 in 28 |
|     in the UK | 1 in 50 |
| That a US President will be | |
|     killed in the next twelve months | 1 in 50 |
| That you will be assaulted | |
|     in the next year (in the USA) | 1 in 100 |
| That you will smoke yourself | |
|     to death in the next year | |
|     (twenty cigarettes a day) | 1 in 200 |

| | |
|---|---|
| That you will enter a mental hospital in the next year | 1 in 333 |
| That you will be killed during the next year on a motorbicycle | 1 in 500 |
| That you will get a full house in your next poker hand | 1 in 694 |
| That you will get killed in the next year driving a car | 1 in 4000 |
| That a baby will die at birth | 1 in 12,500 |
| That you will drink yourself to death during the next year (one bottle of wine a day) | 1 in 13,000 |
| That you will die during a surgical operation | 1 in 40,000 |
| That you will die from taking contraceptive pills (women only) | 1 in 50,000 |
| That you will die in an aircrash in the next year | 1 in 100,000 |
| The chances of a US soldier in Vietnam getting killed | 1 in 125,000 per hour |
| That you will die in the next year from electrocution | 1 in 160,000 |
| That you will get a royal flush in your next poker hand | 1 in 649,740 |
| The chances of a British soldier in Belfast getting killed | 1 in 833,333 per hour |
| That you will be struck by lightning in the next year | 1 in 2,000,000 |
| That you will win the football pools | 1 in 22,000,000 |

A couple of things stand out from this table. First, it is more dangerous to smoke twenty cigarettes a day or drink a bottle of wine a day than it is to fly in commercial aircraft: yet airlines have to hold several treatment clinics to help people over their flying phobias. The second point is that, although the events at the bottom of the table are – theoretically – very rare, no one would deny that air crashes happen, or that soldiers have been killed in Vietnam or Belfast. Rare events do happen.

Our attitudes to risk are stranger than might at first appear. The figures in the table show that you are far more likely to die in a car than in an air crash (twenty-five times more likely in fact). Yet few people are as frightened of cars as they are of aeroplanes. Many examples of this kind could be given, but here is another, quite different, instance. Psychologist John Cohen, from the University of Manchester, showed people a screen divided into four quadrants and asked them to guess in which quadrant a signal would soon be shown. Rather than guess randomly, people chose the quadrants in the following proportions:

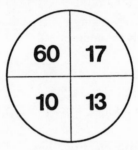

This is intriguing because it shows how curious and irrational people are in their attitudes to chance events. Indeed gamblers can make a lot of money out of the way other people misunderstand probabilities. Ask a series of people to guess any number from 1 to 10 and challenge someone else that you can guess their choice better than 1 in 10. You will win. The great majority of people choose the five numbers between 3 and 7. So if you stick to those, you will guess right nearly five times out of ten – and win your bet.

## Less than rare events

A second point to bear in mind is that some events which seem uncommon are actually not rare at all. The classical example of this is what mathematicians call 'the birthday problem'.

You are at a party. What are the chances that two people in the room have the same birthday? Most people think that this is a fairly unlikely event; but in fact, even if there are only ten people at the party, the chances are one in nine that two of you have the same birthday: if there are 25 people, then the chances are *better than evens* that two of you have the same birthday.

Start with yourself: you have a birthday. So the probability that anybody else in the room has a different birthday is 364/365 – or the rest of the days in the year that are not your birthday divided by the number of days in the year (ignore leap years for the sake of argument). After we cross that second person off we can then say that there are 363 other days left for a third person to have a birthday on. So the probability that his birthday differs from your own and from the second person's is 363/365. These are separate events – they are not linked in any way – so the compound probability that number two differs from you, and that number three differs from both of you is:

$$\frac{364}{365} \times \frac{363}{365}$$

You can go on like this to include all of the people in the party and we can work out the chances that any two people have the same birthday from the formula:

$$1 - \frac{365 \times 364 \times 363 \times 362 \times 361}{365 \times 365 \times 365 \times 365 \times 365} \ldots$$

The multiplications give us the chances that all the birthdays are *different*. We subtract this from absolute certainty, 1, to get the probability that two are the same. The calculation continues until all people in the party are accounted for. The figures work out that, for ten people, the probability is 0.117 (better than one in nine), for 22 people 0.476 and for 23 people 0.507 – or *better than evens*.

First, let us note that these calculations work out in practice as well as in theory. For instance, in the succession of US Presidents you have only to go as far as the 29th before you find a coincidence of birthdays: Warren Harding shared 2 November with James Polk. For deaths (theoretically the same mathematical phenomenon) the coincidences are even more striking: the second, third and fifth Presidents, John Adams, Thomas Jefferson and James Monroe, all died on 4 July.

This is a particularly interesting problem because it is related to one of the striking coincidences in the Minnesota study: that between Terry Connolly and Margaret Richardson, who were married on the same day. Remember that they were the 16th pair of twins seen by Bouchard and his colleagues, that is, the 31st and 32nd people seen at Minnesota. Of these, four sets of twins were not married. This leaves 24 married twins of whom Terry and Margaret were numbers 21 and 22.

We know, from the 'birthday problem', that the chances are nearly even that two of the Minnesota twins will have got married on the same day of the year. Terry and Margaret also got married in the same year so we need to make a modification to the calculation. Let us say, for the sake of simplicity, that almost everybody who gets married does so between the ages of 17 and 37 (which is very nearly true anyway). It follows that the figures mentioned need to be multiplied by 20 to take this into account. This works out as 0.025 or 1 in 40. So there is 1 chance in 40 that two of the twins who have been to Minnesota will have been married on the same day. (The chances are actually greater than this because of the strong tendency in Britain for people to be married on Saturday, and because there used to be tax advantages to early spring weddings.)

But of course, it was not just any two twins who were married on the same day but the two members of a single pair of twins. Can we work out the odds of that happening?

If we say, as above, that almost everybody who gets married does so between the ages of 17 and 37, the chances of marrying in the same year is 1 in 20. But we also know that most people get married in their twenties, so this brings the odds down to, say, 1 in 15. We also know that in Britain most people get married on a Saturday, and that, if you exclude the Christmas holiday period, there are fifty Saturdays in a year. Let us assume that as many people are married on a Saturday as on all the other days of the week put together. That would mean that the chances of the two twins being married on the same Saturday in the year would be roughly 1 in 100 and the chance that they were married on the same Saturday in the same year: 100 × 15 = 1500, 1 in 1500. Since we know that there were twelve pairs of married twins who have been to Minnesota, it follows that the chance that any one of those pairs were married on the same Saturday is 1500/12 = 1 in 125. Thus we should find the fact that Terry Connolly and Margaret Richardson got married on the same day (it *was* a Saturday) no more surprising than the possibility that the reader of this book, if he or she is American, will be assaulted during the next year. (And only twice as unlikely as that the President will be killed in the next year.)

## The perils of numerology

It is easy to be trapped by coincidences into seeing patterns where none in fact exist. All mathematicians are taught the folly of numerology – the bogus power of numbers, and in this field we do well to remember it.

A good example comes from the Second World War. In 1944 someone discovered intriguing patterns in the lives of the world's leaders. This pattern can be set out as follows:

|  | Year of birth | | Age | | Year occupied office | | Years in office | |  |
|---|---|---|---|---|---|---|---|---|---|
| Churchill | 1874 | + | 70 | + | 1940 | + | 4 | = | 3888 |
| Hitler | 1889 | + | 55 | + | 1933 | + | 11 | = | 3888 |
| Mussolini | 1883 | + | 61 | + | 1921 | + | 23 | = | 3888 |
| Roosevelt | 1882 | + | 62 | + | 1933 | + | 11 | = | 3888 |
| Stalin | 1897 | + | 65 | + | 1924 | + | 20 | = | 3888 |

The unchanging grand total, 3888, was twice the year: 1944. Was there a lesson hidden in these figures pointing to some important conclusion? Would war end in 1944? And if the year of the war's end could be found by dividing the magic number by two, could the month, day and hour of peace be predicted by dividing in half again? Half 1944 is 972 – representing, so it was thought, the ninth month (September), the seventh day, at 2 a.m. They were wrong.

(They *had* to be wrong, for this carefully constructed table is, of course, a con. The formal proof of this may be seen if we re-write the table as follows:

| | Year of birth | Age | Year occupied office | Years in office |
|---|---|---|---|---|
| Leader | $(A)$ + | $(1944 - A)$ + | $(B)$ + | $(1944 - B)$ |

A moment's inspection will show that the answer in all cases has to be $1944 + 1944 = 3888$.)

Probably the most absurd use of numerology to produce a pseudoscience was that started by Dr Wilhelm Fliess, a Berlin surgeon. Fliess was obsessed with two numbers: 23 and 28. He was convinced (and, in his time, managed to convince quite a few others) that two fundamental cycles underlay everything in nature: a male one of 23 days and a female one of 28 days. The two cycles, he said, were present in every living cell; by manipulating these two numbers – multiplying at times, subtracting or adding at others – he sought to explain a great deal. For example, Fliess explained that the age of 51 'seems to be a particularly dangerous one' because it was the sum of 23 and 28. Fliess became a great friend of Sigmund Freud, the psychoanalyst, and for years Freud expected to die at the age of 51. Fliess thought the different cycles explained illnesses like neurasthenia and anxiety neurosis. His books, of which he wrote a great many, list such things as:

The cycle of the moon around the earth – 28 days
A complete sunspot cycle – 23 days
The content of dreams – recurs every 23 days
Accident-proneness – people tend to have 'off' periods every 11½ days, half of 23

Fliess's theories were nonsense and he was rightly discredited. Freud retained his feeling for the man, but even he realized that Fliess's numerology was absurd. I refer to it to emphasize that patterns or rhythms in coincidences do not necessarily add up to anything.

## The natural history of coincidences

Arthur Koestler is a Hungarian by birth though he has lived in Britain now for many years. He has written several novels, half-a-dozen books of autobiography and at least a further dozen essays on scientific and artistic subjects. He has also been an avid collector of coincidences, and he makes three points that we would do well to bear in mind in this chapter.

The first is that a great many coincidences involve names of one kind or another. The most notorious, according to Koestler, is the case of Richard Parker. In 1838 Edgar Allan Poe, the American horror story writer, wrote a story called *The Narrative of Arthur Gordon Pym of Nantucket*. In that story Mr Pym stows away in a ship that is wrecked. There are four survivors, who spend many days in an open boat before they are driven to kill and eat one of their number, an ordinary seaman called Richard Parker.

In 1884 – forty-six years later – the yawl *Mignonette* foundered and the four survivors were in an open boat for many days. Eventually the three senior members of the crew killed and ate the cabin boy. The cabin boy's name was Richard Parker.

The British House of Commons provides a striking coincidence of names at present. Among the 630 MPs are not one but two pairs with the same name: one Ron Brown represents the constituency of Leith, another Ron Brown was elected for Hackney South and Shoreditch; the former Prime Minister Jim Callaghan, MP for Cardiff South East, is 'shadowed' by another Jim Callaghan, who sits for Middleton and Prestwich. And although Labour MPs make up much less than half the House, all four 'doubles' are Labour members.

It is not clear why names should feature in so many co-incidences but maybe it is because we all have them, they are part of our individuality and therefore we are more alert to episodes involving them than anything else. As is already apparent, coincidences among names are a chief feature of the Minneapolis study.

We should keep in mind, however, that there are plenty of differences between the 'twin' MPs. All four, for instance, entered the House at different elections. One Jim Callaghan has been Party Leader and Prime Minister, the other has always been a backbencher. One Ron Brown is on the right of the Party, the other is on the left. We could go on, of course, listing differences, whereas the similarities are limited in number – though the number may increase if we search hard enough for them. There is danger in the 'numerologist's disease' – of looking for whatever pattern there might be and then letting it dominate the picture and appear to 'mean something'.

Similarly with the Richard Parkers. Maybe the people on the *Mignonette* had read Poe's book. Then, after the disaster occurred, they could not help but be aware of the coincidence of even having a Richard Parker aboard (it is not such an unusual name). Then, perhaps with an eye to the box office should some of them get safely home, they actually settled on Parker for that reason. Who knows? There were many other differences, too – in the ship, its route, in the ranks of the two Parkers (one was a cabin boy, the other an ordinary seaman), and in the sequence of events (a mutiny takes place in the Poe story, not in the sequel). And so on. For the moment, though, we will stick with Koestler's point that a lot of coincidences involve names.

His second point is that many other coincidences involve numbers. The most celebrated of his cases involved a Mr Peter Moscardi from Essex near London (and how similar that unusual name is to one of the twins in the Minneapolis study – Helen Moscardini).

Moscardi, a constable in the Metropolitan Police, was stationed in 1967 at a suburban police station on the edge of London. One day that autumn a friend said he had tried to contact Moscardi at the station but was unable to reach him. The reason was simple – the telephone number had been changed a day or so before. Moscardi explained what had happened and gave the friend the new number – 40166. The friend said he would be in touch.

Later, Moscardi realized he had given his friend a wrong number: the correct one was in fact 40116. The friend did not have a phone, so Moscardi was unable to do anything about it. On night duty a few days later he was patrolling an industrial estate at about 11.30, when he noticed that the front door to the manager's office was open. He went to investigate but, as he entered the office, the telephone started to ring. He lifted the receiver and asked what number the caller was trying to reach. In Moscardi's words: 'A voice, which I realized was my friend's, asked to speak to me. Mystified I looked at the dial but there was no number shown. I subsequently learned from the manager of the premises that his private telephone, which was unlisted, was 40166 – the same as the incorrect number I had given my friend.'

This anecdote also illustrates the third point that Koestler makes, namely that some coincidences are not only unusual events but they also seem to have a purpose to them. Mostly, he says, this purpose appears to be to return lost property to people. One nice example involves the film star Anthony Hopkins.

In September 1971 George Feifer, an American writer living in London, lent a friend a copy of his own novel, *The Girl from Petrovka*. It was a personal copy, full of notes in the margin, but the friend lost it when it was taken from his car in Bayswater. Two years later, Feifer travelled to Vienna for the novel's filming and met Mr Hopkins, who was to star in the movie version. There he was struck by a strange tale Mr Hopkins told him.

Having liked the script and signed to appear in the film, the star had gone to central London to buy a copy of the book. He could not find one but, on the way home, he came across a packet on a bench in Leicester Square underground station. At first he thought it was a bomb but when he turned it over he saw that it was a copy of *The Girl from Petrovka*. He was even more intrigued by all the red marks in the margin. 'They wouldn't have any personal significance for you, would they?' he asked George Feifer.

One has to admit, I think, that whereas some of the

coincidences noted by Koestler are clearly little more than chance, these last two cases are difficult to explain – or to explain away.

So with that tantalizing thought, let us now move on to the Minnesota results, and see how they fit into the general body of coincidences.

## The coincidences at Minneapolis

Fundamentally, we need to look at two things. In the first place, we need to take some of the more striking coincidences and try to work out exactly how likely or unlikely they are. Second, we need to have some idea of the chances of two people having more than one set of coincidences in common. For instance, the two Jims named their children the same *and* had the same pattern of headaches *and* went to the same beach for a holiday. One coincidence may not be all that rare – but several may be.

First then, let us look at some of the more striking coincidences. We can at this stage exclude things which have any link with physical characteristics – headache, poor eyesight, patterns of hair loss and so on. As we saw from the last chapter, these are part of the natural bond – there is nothing uncanny about them. We shall concentrate instead on the other characteristics which twins share. I have divided coincidences into those for which we have some idea of their distribution throughout the general population, those for which we have no such knowledge but for which knowledge could be acquired, and those which are at present too vague to be sensibly measured.

### Coincidences which can be measured
The distribution of names:
    for wives or husbands
    for children
    for pets
Jobs
Hobbies and mannerisms (like biting fingernails)
Favourite:
    subject at school

colour
drink (hard and soft)
author
newspaper and magazine
food
Clean-shaven versus moustache or beard
Musical habits
Marital status
Number of children, order, ages and spacing
Fears

*Coincidences which can be measured but which are not properly quantified at the moment (all these occurred in the Minnesota studies)*

Holiday places
Dreams
Favourite clothes
Frequency of:
   drinking (specific drinks)
   eating (specific foods)
   giggling
Falling asleep in front of the television
Eating alone in restaurants
Reading magazines back to front
Flushing the lavatory before use
Circumstances of meeting spouse
Date of wedding

*Coincidences too vague or complicated to measure*

Many verbal expressions
Many gestures (like gait or way of sitting)
Habits (like putting off things until tomorrow)
Sense of humour

This table excludes the difficult matter of timing – such as the chances of people wearing their favourite clothes on a particular day.

We will now turn to some of the characteristics in the first two parts of the table and try to work out how unusual the coincidences between identical twins reared apart really are.

*Fears*

Not surprisingly, perhaps, quite a bit is known about our hang-ups. Surveys show that three out of every five Americans admit to at least one fear, with women much more affected than men (75 to 38 per cent respectively). Here are some of the common fears reported by the polling institutions:

|  | Percent having a specific fear | |
|---|---|---|
|  | *Men* | *Women* |
| Speaking before a group | 36 | 48 |
| Heights | 26 | 38 |
| Deep water | 13 | 30 |
| Loneliness | 11 | 16 |
| Dogs | 9 | 14 |
| Lifts | 4 | 11 |
| Escalators | 2 | 8 |

In addition, another poll, which did not divide respondents according to sex, found that 46 per cent of all people feared snakes (in addition, another 20 per cent admitted they were 'slightly bothered'), and that 1 million Americans are agoraphobics – that is, have a fear of going out into wide open spaces. (You can get a good idea of the size of the problem when you realize that 1 million people represents only 0.5 per cent of the US population: compare that figure with the percentages for other fears in the table above.)

Fears show some of the problems we have in dealing with statistics. In some cases we have 'global' statistics, applying to whole populations; some divide according to men and women; and others, as we shall see in the case of nailbiters, are broken down according to age.

Several of the twins had a fear of heights – mainly women like the British twins, Jean and Irene, who did not like Bouchard's car ride up to the top of the multistorey car park. We can see from the table above that the chance of any woman having this fear is 38 per cent or, as it is usually written by statisticians, 0.38 (on the basis that absolute certainty is represented as 1). Therefore, the chance of any *other*, unrelated person sharing that fear is 0.38. In other words, there is 1 chance in 2½ that *anybody taken at random* will share a fear of heights with Irene and Jean. And this is, of course, no surprise at all. Horses that start races with these odds are winning all the time.

Now let us turn to the fear of deep water. The relevant figure here, expressed as a probability, is 0.30 for women. Applying the same calculation as before, we can work out that the chance of any twin sharing this fear is about 1 in 3; slightly less than the fear of heights but hardly surprising none the less.

If we take agoraphobia – one of the rarer and more inconvenient fears – we still find that 1 person in 200 has this fear. This is a much smaller proportion than for fear of deep water, but then again it is 250 times *as common* as Sean Connery's win at the roulette table, or as being killed on a motorbike in any one year.

The general picture so far as fears are concerned, therefore, is that there is nothing remarkable in identical twins sharing so many. The fears are so common (with the exception of agoraphobia) that almost any two people drawn at random from the street would share some of them.

*Nailbiters*

We have seen that Oskar Stohr and Jack Yufe both bit their nails down to the quick. According to Alexander Goldenburg, a dentist at the Mount Sinai Hospital in New York, 24 per cent of Americans bite their nails, but this varies according to age:

| | |
|---|---|
| under 10 | less than 3 per cent |
| 21–30 | less than 61 per cent |
| 30 and above | 14 per cent |

So the chances of any person over 30 years old being a nailbiter is 14 per cent or nearly 1 in 7. Assume for the moment that only a third of nailbiters bite all the way to the quick. For the over 30s the figure is now roughly 5 per cent – 1 in 20, roughly the same as the risk you face in the next year of having your home broken into.

*Bedwetters*

Problems such as enuresis and stuttering may have a physical or genetic base, in which case coincidence between the twins on this count should not be at all surprising. However, figures are available on the incidence of these disorders, so we can still work out the probabilities of their joint occurrence in any two, quite unrelated people.

According to a 1960 US Department of Health, Education and Welfare study, 5 per cent of people in their teens wet their beds. But the Minnesota twins were asked, not if they wet their beds at the time of the study or in their teens, but if they had *ever* wet their beds (which many people do as young children). The figures that follow therefore are likely to be an underestimate. A 5 per cent probability for any one individual translates, in our formula, into 1 chance in 20. Unusual but not outstandingly so, especially in view of the fact that it is almost certainly an underestimate.

*Stuttering*

About 12 million Americans have some kind of speech defect, mostly speaking with a lisp. Serious stutterers run to 1.5 million, or 0.68 per cent of the population. Now we are getting to rarer aspects of behaviour: 0.68 per cent means that only 1 person in 147 will share this disability with a twin who has it. Even so, that is nearly five times *as common* as getting a full house at poker.

*Neatness*

Two of the British female twins in the Minnesota study (Bridget and Dorothy) were compulsively neat at home. Statistics show that women are more bothered by untidiness to begin with but get less fussy as they grow older. Men, on the other hand, show the opposite tendency. Roughly speaking, however, the figures indicate that about 7 of every 10 women say that they feel uncomfortable when they are in a house that is not completely clean and tidy. Let us assume that only one-tenth of them could be called 'compulsively' tidy: always shifting things around, cleaning when it is unnecessary, and so on. This gives us a figure of 7 per cent who fit the criterion, that is 1 in 14. This condition is less rare than you would think.

*Falling downstairs*

The figures in this section are necessarily crude and our answers can be no more than an educated guess. But we have seen that both Daphne Goodship and Barbara Harrison fell downstairs at the age of fifteen and ended up with weak ankles as a result. Let us see how near we can get to calculating the probability of this coincidence.

According to figures released in 1976 there are, in America, 538,500 injuries a year as a result of falling downstairs. Now there are four sets of leg joints, so let us assume that only a quarter of these falls result in a damaged ankle (as opposed to toes or knees or hips). Disregarding those clumsy people who make a habit of this and fall down the stairs more than once a year, that makes 134,625 people to base our calculations on. The probability of any single person falling down stairs in a particular year is, therefore:

$$\frac{134,625}{220,000,000} = 0.0006$$

or 1 in 1666. On the other hand that is 60 times more likely than dying in an air crash during the next year – and there is a major air crash, on average, every week somewhere in the world.

*Eating and drinking*

Although a lot is known about our eating habits, it is not always possible to put a precise figure on any particular activity. Here, for instance, is a list of the items that sell best in (American) restaurants:

Appetizer: shrimp cocktail
Soup: vegetable
Sandwich: hamburger
Entrée (meat): roast beef
Entrée (non-meat): spaghetti
Entrée (fish): fried shrimp
Side dish: green beans

Dessert: apple pie
Juice: orange
Salad dressing: French

As you see, there are no figures attached but clearly for any two people, twins or otherwise, to share one of these favourites is no surprise at all. More specifically, American restaurant figures show that when people eat out they order coffee 59 per cent of the time at breakfast, 20 per cent of the time at lunch, and 16 per cent of the time at dinner. Now we saw that Daphne and Barbara both listed black coffee, with no sugar, as their favourite drink. Let us be very conservative and assume, from the figures above, that 59 per cent of people, and no more, drink coffee at some point in the day. Let us assume further that half of those coffees are black – 29.5 per cent – and that half of those – 14.75 per cent – are without sugar. That gives us a (rounded) probability for a black coffee (no sugar) drinker of 0.15 or 1 in 6.6.

The picture with hard drink is also patchy. We know, for instance, that the favourite brands of drink in America are:

Scotch: Chivas Regal
Rye: Seagram's Seven
Bourbon: Jack Daniel's
Gin: Beefeater
Vodka: Smirnoff
Rum: Bacardi
Liqueur: Kahlua

But we do not have figures to attach to them. On the other hand, we do know that the percentage of American men and women who drink various liquors is as follows:

|         | Men | Women |
|---------|-----|-------|
| Vodka   | 12  | 11    |
| Bourbon | 12  | 6     |
| Scotch  | 11  | 5     |
| Gin     | 8   | 6     |
| Brandy  | 5   | 4     |
| Rye     | 6   | 4     |
| Liqueur | 4   | 4     |

The British twins, Barbara and Daphne, who invented the cocktail 'Twin Sin', based it on vodka, which was, in both cases, their favourite drink. We can now see that vodka is the most popular drink among women, with a probability of 0.11, 1 chance in 9, that any woman will drink it. It is scarcely odd, therefore, that both Daphne and Barbara like vodka. We can go so far as to say that, if they are going to share their drinking habits, it is more likely to be vodka than anything else.

However, Oskar or Jack (we don't know which), and Gladys and Goldie, too, were not just drinkers but hard drinkers, 'alcoholics' at times. According to the US Department of Health, Education and Welfare, anybody who drinks one alcoholic beverage a month is a 'drinker' but they break down drinkers into light, moderate and heavy. The dividing line between moderate and heavy is one ounce of absolute alcohol per day. This is equivalent to any of the following: two shots of straight whisky; two four-ounce glasses of wine; two beers. The breakdown of drinking habits is as follows (remember that on average people understate their drinking by about a half):

|  | Men % | Women % |
| --- | --- | --- |
| Abstainers | 23 | 44 |
| Light drinkers | 33 | 36 |
| Moderate drinkers | 26 | 14 |
| Heavy drinkers | 16 | 6 |

On this basis alone, therefore, the fact that *both* Oskar or Jack and Gladys and Goldie were heavy drinkers was not too surprising. For men the relevant figure is 16 per cent or 1 in 6. For the women 6 per cent equals 1 in 17.

On the other hand, there are several qualifications which need to be made to these figures. Heavy drinking takes place most among people in their twenties; Jews show a smaller proportion of abstainers than Catholics or Protestants but also have the lowest proportion of heavy drinkers. This latter fact, particularly, could affect the odds in our case, as could the different consumption patterns of the USA and Germany (Jack Yufe is American Jewish, Oskar Stohr German Catholic).

*Comparative alcohol consumption by countries*

|  |  | Litres per head per annum |
| --- | --- | --- |
| *Spirits* | Peru | 12.5 |
|  | Soviet Union | 12.5 |
|  | USA | 10.0 |
|  | Canada | 9.5 |
|  | W. Germany | 9.1 |
| *Beer* | W. Germany | 191.5 |
|  | Czechoslovakia | 188.5 |
|  | Belgium | 185.5 |
|  | Australia | 182.0 |
|  | New Zealand | 177.5 |
|  | USA | 109.0 |
| *Pure absolute alcohol* | Portugal | 23.5 |
|  | France | 22.5 |
|  | W. Germany | 15.0 |
|  | Belgium | 14.4 |
|  | Austria | 14.0 |
|  | USA | 10.5 |

Gladys and Goldie had both been classified as alcoholic. What are the chances of this arising by accident? The figures on alcoholism vary according to who is doing the survey and what criteria are used (e.g. deaths from cirrhosis of the liver, or the amount of alcohol consumed). In America the overall figure varies from 5.75 million (3.3 per cent of the population) to 10 million (5.7 per cent), with women numbering from 16 per cent to 50 per cent of the total. Taking an average, we can say that 2.5 per cent of the population are alcoholic men and 1.5 per cent are women (making 4 per cent overall, which is between the two extremes, 3.3 and 5.7). Therefore there is 1 chance in 40 that *any* man will be alcoholic, and 1 chance in 67 that *any* woman will be. This is a much greater probability than the chances that you will be assaulted in the next year.

## Shyness

More than one pair of twins reported being chronically shy. Shyness was explored by psychologist Philip Zimbardo, in 1978. In his book, *Shyness*, he gives the following percentages:

70 per cent of all those sampled said they were shy
40 per cent said they were chronically shy
2 per cent said they were shy all the time, wherever they were and whoever they were with

Two calculations are therefore possible – using either the chronically shy figures of 40 per cent or the 2 per cent figure representing what we might call the pathologically shy. As probabilities this comes out as:

0.4, or 1 in 2½ for chronically shy
0.02, or 1 in 50 for pathologically shy

The chances are that the twins were not particularly shy and were merely examples of a trait that is far more widespread than we think.

## Jewellery

One of the most extraordinary coincidences in the Minnesota study was that between Dorothy Lowe and Bridget Harrison; both arrived at Minneapolis wearing three rings on one hand and a bracelet and a watch on the other. Discounting the watch – since most people wear them – what are the chances that a woman will wear rings and bracelets?

According to *Jeweler's Circular Keystone*, 40 per cent of Americans over 18 buy at least one piece of jewellery a year. The figures do not tell us about jewellery which is inherited, or whether they buy it for themselves or for others, nor to what extent the 40 per cent who buy in one year are the same 40 per cent who bought jewellery the year before. But these figures show that jewellery-buying is common, and that it is not perverse for us to assume, for the sake of argument, that people on average buy a piece of jewellery every three years (based on the premise that 33 per cent of the population buys something every year, with no one buying in between times). This would mean that, between the ages of 18 and 35, Bridget and Dorothy would have bought five or six pieces. Further, their husbands would have bought the same amount. Guessing, we could say that most of what the husbands bought was for their wives. This would give Bridget and Dorothy each about ten pieces of jewellery.

The jewellery trade also tells us that the breakdown of jewellery sold, at least for diamonds, is as follows:

Rings – 60 per cent
Necklaces and pendants – 22
Earrings – 12
Pins – 2
Bracelets – 1.5

Let us assume that the proportions of rings, bracelets, etc., is the same for other precious stones as it is for diamonds. (Strictly, this is not likely to be true: diamond rings are popular probably because they use only a few stones whereas bracelets use many. Cheaper stones may feature more as bracelets for this reason. However, this breakdown is the only one we have to go on.) It follows from these figures that, of the ten or so pieces which Bridget and Dorothy would have, six would have been rings, two would have been necklaces and one or two would have been bracelets. One wonders therefore what all the fuss was about. This pair of twins travelled to Minnesota together and their reunion had occurred some time before. Who can tell what subtle influences were at work to make them dress similarly? And even if there were no such influences, is it really so odd that they each wore three of their rings on the same hand and a bracelet on the same wrist? They only had two hands and two wrists each, after all. Certainly in my experience many women wear several rings on their wedding fingers, quite apart from their wedding ring.

*Dress*

In 1978 a *Women's Wear Daily* survey showed that the average woman in the United States had the following wardrobe:

Panties: about 12 pairs (with an average
   replacement of 5 per year)
Slips: 3
Robes: 3
Nightgowns: 5
Blouses: 10
Skirts: 8
Daytime dresses: 5
Cocktail dresses: 3
Jackets or blazers: 2
Coats and heavy jackets: 3
Raincoat: 1
Slacks: 7
Sweaters: 5
Jeans: 2

The twins who turned up at their reunion dressed in the same way were wearing a skirt, blouse and sweater. Leaving colour aside for the moment we can see they might each have had the following outer clothing to choose from:

Blouses: 10
Skirts: 8
Daytime dresses: 5
Jackets or blazers: 2
Raincoat: 1
Slacks: 7
Sweaters: 5
Jeans: 2

A jacket, blazer, coat or raincoat would probably have been worn over the rest, so in the end we are dealing with just blouses, skirts, dresses, slacks, sweaters and jeans as the average wardrobe the twins had to choose from – 37 items in all. Of these, five were dresses. When a dress is not worn, jeans, skirts, or slacks with blouses and/or sweaters take its place. Assuming throughout, therefore, that all clothes have an equal chance of being worn, a woman will either wear a dress or a skirt or slacks or jeans in these proportions: dresses 5; skirts 8; slacks 7; jeans 2. Or, since this totals 22 garments, dresses 5/22, skirts 8/22, slacks 7/22 and jeans 2/22.

The chance of a woman wearing slacks therefore is 7/22, or 0.31, and the chance of her wearing any particular set of slacks is 1/22 or 0.04. Assuming that all her blouses match all her slacks equally, the chance of wearing a particular blouse (she has 10) with this particular pair of slacks is ten times as unlikely, which is 0.004. Once again, assuming that all her sweaters match all combinations of slacks and blouses equally this probability is five more times as unlikely still. The calculation works out as 0.0008, or 1 in 1250.

In none of the cases in which the twins wore similar clothing was it claimed that the arrangements were identical – just similar. So we do not need to take account of the hundreds of different styles of blouses, skirts and slacks.

Next take colour. If we assume seven main colours – black, white, blue, green, yellow. red and brown – we can note first that this is the same as the average number of slacks or blouses a woman has and only just more than the average number of sweaters (5). Assuming that all sweaters, blouses and slacks are different colours therefore, the odds should not be much affected. In fact they are probably reduced. Most of us have similar ideas about what colours match (pale pink with dark red, green and yellow, etc.) and so it follows that selecting a blue skirt, say, limits the number of blouses that can be worn with it, perhaps by a factor of 2 or 3. The chances therefore of *any* woman being dressed in a very similar way to Dorothy or Bridget seem to run roughly at 1 in 1250. This is fairly rare, of course. On the other hand it is eight times more probable than being killed in a plane crash in the coming year. However needless these deaths, does anyone think them surprising? Or deny that they occur? Or think that they have a supernatural cause?

## Smoking and homosexuality

I have put these together not because they are linked in any psychological or medical sense but simply because, in both cases, high rates of 'discordance' have been found in twin studies, when perhaps substantial rates of coincidence might have been expected. That is to say, in many cases one twin smokes but the other does not; one twin is homosexual, the other is not. At the same time, spectacular examples of coincidence are also common: in the Minneapolis study the most striking is that of the homosexual twins who now live, and sleep, with each other. (In twin studies of smoking, concordance rates for MZ twins range from 50 to 70 per cent and for DZ twins from 24 to 69 per cent.)

Nearly one in three American adults smokes cigarettes, 54 million people, plus another 6 million who smoke cigars and pipes. Men, it seems, are smoking less, but many more women are lighting up – an increase of 6 million since 1970. The actual figure, based on an adult population of 154 million, is 35 per cent, or slightly less than 1 in 3. Heavy smokers – more than thirty-five a day – make up 17 per cent of all male smokers (4,590,000) and 10 per cent of women (2,700,000). For the total adult American population the percentages are 5.96 for men and 3.5 for women – that is, 1 man in 7 and 1 woman in 30, or thereabouts. So, even for heavy smoking, the odds are not terribly long: ten times as common as a full house in poker.

Homosexuality has only cropped up in the Minneapolis study among the men. In one pair both were homosexual; in another, only one brother was. In America the Institute for Sex Research estimates that 6 per cent of men are actively homosexual. This is an estimate, it must be understood, as homosexuality is one of those issues on which it is notoriously difficult to get a true picture through survey research. If the estimate is right the chances of *any* man sharing this trait of homosexuality with a twin is roughly 1 in 16: not at all unusual. Studies on homosexuality among twins have sometimes shown high concordance, but only in one study was the number high enough for us to attach much credibility to it. In that, all 37 MZ twin pairs were concordant, whereas only 12 per cent of 26 DZ pairs were.

*Piano playing*

Two of the twins, Dorothy and Bridget, played the piano. In 1978, according to Gallup, about 18 million Americans did so, 79 per cent of them – 14,220,000 – being women. Let us assume that no one under the age of five plays the piano (many children, after all, start to learn very early on). In America the population over that age is 203,500,000. So the percentage of women playing the piano is:

$$\frac{14,220,000}{\frac{1}{2} \times 203,500,000} \times 100$$

which equals 13.9 per cent. So *any* woman we pick at random stands a 1 in 7 chance of playing the piano. It's more common than you might think.

But these two twins also gave up the piano after the same grade. Without knowing the proportion of people who give up, we cannot fully work out this likelihood. However, we do know that, on average, 28 per cent of Americans in their late teens and twenties play a musical instrument but that by the time they are in their sixties only 10 per cent do. This is a reduction of 4.5 per cent (or 16 per cent of players) every ten years. That is, of those who play, about 1.6 per cent stop in any year. Since grades in Britain broadly equate to years (at least so far as schoolchildren are concerned), we can say that the probability that any person who used to play the piano and stopped in the same year as one of the twins, purely by chance, is:

$$0.14 \times 0.016 = 0.0003 = 1 \text{ in } 3333$$

– slightly more common than the chance of being killed next year in a car crash.

## Cars

It will be remembered that both Jims drove to the same beach spot for their holidays – and that both went in a Chevrolet. According to figures released by the American Motor Traders Association, the Chevrolet was the best-selling car in 1978 and, assuming that foreign cars accounted for one-fifth of all cars sold, the proportion of cars on the road at that time which were Chevvies was 14.4 per cent. So the chances of anybody driving the same car as one of the Jim twins was only 1 in 7.

Where is all this pointing? I would suggest as follows: a few of the coincidences in the Minneapolis study are, when you look at them in detail, not as rare as you may have thought. Others which are, statistically speaking, truly rare, are nevertheless no rarer than events which are taking place all the time – car and plane crashes, deaths due to war and so on, which no one attributes to 'uncanny' or 'spooky' reasons. It is perhaps surprising that these coincidences are, statistically, not surprising.

## The roots of coincidence

But of course the Minnesota twins did not have just one coincidence in common; each pair had several at least. This is more difficult to analyse, since we do not know how the coincidences are linked. For instance, both the Jim twins went to the same stretch of Florida beach in a Chevrolet. Now it could be, though we cannot be certain, that people who drive Chevrolets do so because they earn less than average and the Chevrolet is a cheap car. Similarly, they holiday in Florida because it is a cheap state, and they drive there because it is cheaper than flying. In this case the coincidences may owe something to the fact that both Jims are, to say the least, not millionaires. And we have already seen that constitution can affect lifestyle. Similarly, the twins who gave up playing the piano may have done so because of some other change in their life – illness, for instance, or having a baby; in other words, several coincidences which appear to be separate may all stem, directly or indirectly, from a single coincidence, rather in the way several diverse symptoms can signal one illness.

There are two things we can do, faced with this problem. We can simply assume that the coincidences are all unrelated, or we can try to work out how a chain of coincidences might have arisen from a common source. To begin with let us assume they are all independent. How does that affect the probabilities of the Minnesota twins?

## The Jim twins

| | |
|---|---|
| Both drive Chevrolets | 0.14 |
| Both bite fingernails | 0.05 |
| Both heavy smokers | 0.059 |
| Both heavy drinkers | 0.16 |

If these coincidences are independent, the chances that these four could arise by accident are:

$$0.14 \times 0.05 \times 0.059 \times 0.16 = 0.00006608$$

or 1 in 15,152, ten times as likely as being electrocuted during the next year.

## Barbara Herbert and Daphne Goodship

| | |
|---|---|
| Both fell downstairs aged 15 | 0.0006 |
| Both share a fear of heights | 0.38 |
| Both drink vodka | 0.11 |
| Both drink black coffee | 0.15 |
| Both wear similar clothes | 0.0008 |

Once again, assuming all these coincidences are independent, the chances of these all occurring by accident are these five fractions multiplied together, which in this case yields: 0.000, 000,003,009,60, or 1 in 333 million, 15 times as unlikely as winning the pools.

These are extremely small probabilities indeed, and by themselves they seem to suggest that there is something between twins which makes them susceptible to some form of coincidence. And we have to remember that we have not considered by any means all the striking coincidences unearthed at Minneapolis. However, before we can accept the argument that there is something 'special' between twins, there are several other objections to be looked at.

In the first place, let us assume, in the three pairs of twins above, that other factors reduce the rarity of the coincidences. We have already seen that, in the case of the Jims, their lack of earning power, which *could* stem from a limited intellectual ability, may help increase the chances that they both drive the same car, holiday in the same state, and so on. Suppose, because of this, that in two coincidences between the Jims the odds are reduced by a factor of three. This would then decrease the joint probability of the various coincidences for the Jims to:

$$0.42 \times 0.05 \times 0.059 \times 0.48 = 0.00059$$

or 1 in 1666, half the odds that you will be killed in the next year in a car crash.

In the case of Barbara Herbert and Daphne Goodship we could assume that both were born with a similar (inherited) weakness in their ankles, which makes them more likely to fall downstairs. Such a weakness could easily make such an accident twenty times more likely. Both were the same age, appearance, build and colouring, and this could have affected their choice in clothes, so increasing the likelihood that they would wear similar things – let us say three times more likely for age and three times more likely for looks and build – by a factor of nine in all. Similarly, vodka drinking has increased a great deal in the past few years and both could have been part of that trend, again being the same age. That trend could have reduced the odds on vodka drinking by, say, a factor of five. The changed probabilities for Daphne and Barbara therefore become:

$$0.012 \times 0.38 \times 0.55 \times 0.15 \times 0.0072 = 0.000,002,82$$

or 1 in 350,000: still a tiny probability.

Several comments can be made about these figures. The probabilities vary greatly from twin-pair to twin-pair depending on what they have in common. Even with all the allowances, Barbara and Daphne seem to produce coincidences with a very rare degree of probability. The same cannot be said for the Jims (for the things we are able to measure). Their coincidences, especially when we make allowances for the fact that some of them may not be independent of each other, are not so impressive.

Second, we have to take into account the interdependence of some coincidences. We shall come back to this point.

Third, all the twins had other things in common, the probabilities of which we have no way of measuring. The list of these is long: reading magazines back to front, leaving the bedroom door slightly ajar, 'squidging' up one's nose, falling asleep in front of television and so on. Some, like leaving the bedroom door open, must be fairly common; others – 'squidging' – much less so. In either case, of course, the overall probability of any two people sharing a longer list of characteristics must get very small indeed.

But even if it is true that the chances of twins sharing a long list of habits are very small indeed, does that signify anything paranormal? Improbable events *do* occur without the need to assume anything supernatural to account for them. The question remains: just how rare does something have to be for it to totally defy the laws of nature? There is no straightforward answer, of course. In many experiments scientists accept that anything beyond a 1 in 100 chance is probably not due to luck, and anything beyond 1 in a 1000 definitely is not. But those are in controlled circumstances where the element of chance is ruled out by the design of the experiment anyway. In our case this is not enough.

## Parapsychology and twins

Professor J. B. Rhine at Duke University in North Carolina has spent some years investigating parapsychology – known also as extrasensory perception (ESP), precognition or psychokinesis – in a series of experiments which usually involve subjects being shown sets of cards with special designs on them – crosses, wavy lines and such like. The person has either to predict what card is going to come up next or else 'wills' a particular card to appear. The forecasts never work all the time; nor are they expected to. Instead Professor Rhine compares the results obtained over a long series of trials with what should be achieved by chance. Simplified, it is like guessing heads or tails. Anyone who tries should be right about half the time. On the other hand, if, over a long period, someone guesses correctly 60 per cent of the time, either he is psychic or the penny is weighted. Or so Dr Rhine's reasoning goes.

Now some of Rhine's experiments have yielded results so improbable that they would occur by chance less than once in 10,000,000,000 times. However, the mathematician Warren Weaver is not convinced that this automatically means that the paranormal is involved. As he puts it, on the one hand we are asked to believe that a highly improbable chance result has occurred; on the other, 'We are asked to accept an interpretation [the psychic] that destroys the most fundamental ideas and principles on which modern science has been based; we are asked to give up the irreversibility of time, to accept an effect that shows no decay with distance and hence involves "communication" without energy being involved; asked to believe in an "effect" that depends on no known quantities and for which no explanation has been offered, to credit phenomena which are subject to decline or disappearance for unexplained and unexplainable reasons.' Weaver concludes, 'All I can say is, I find this a very tough pair of alternatives.'

Nevertheless on the very next page he admits that he finds 'this a subject that is so intellectually uncomfortable as to be almost painful. I end by concluding that I cannot explain away Professor Rhine's evidence and that I also cannot accept his interpretation.'

What can we add to this? The 'improbabilities' of the twins' coincidences are of a similar order to Rhine's results so we could do worse than adopt Weaver's words by way of conclusion. Recently, however, Lindon Eaves and Krystyna Last, at the University of Birmingham, have looked for a special bond between twins that might account for telepathic or other paranormal occurrences. Eaves and Last examined a small sample of twins, thirty-four pairs in all, but gave them fairly thorough tests. There were ten MZ male, eleven MZ female, five DZ male, three DZ female, and five DZ mixed pairs of twins, aged between fourteen and twenty-two. The twins were put in separate rooms and given a Public Opinion Inventory, together with these instructions: 'We are interested in how well you know your twin's opinions. We would like you to fill in this questionnaire *as you think your twin would*. Please read the instructions and then answer the questions as you think your twin would.' After this the twins were given another copy of the questionnaire to fill out for themselves.

Eaves and Last found that twins rate themselves as *more like* each other than they actually are. Twins *assume* they are alike in spite of evidence to the contrary. Even more important, Eaves and Last could find no evidence that MZ twins could predict their co-twin's responses any better than DZ twins could. In fact, neither sex nor zygosity produced any effect. Finally, Eaves and Last report that twins are not very good at predicting what their co-twin will say about any particular subject; they did hardly better than chance at this prediction – and this is scarcely what would happen if there was any kind of telepathy or paranormal understanding between them.

This result is also supported by a study from Japan, directed by Professor Shoji Watanabe from the Research Institute for Nuclear Medicine and Biology at the University of Hiroshima. Watanabe looked at MZ twins, one of whom had been at Hiroshima and survived the A-bomb attack, the other of whom had not been there at the time. He found that twins were no better than anyone else in understanding the trauma their co-twin had been through, and he used quite a range of psychological tests. His concluding words: 'It [seems] that even between closely related twins the depth mentality of the twin is not readily understood by his counterpart.'

For the moment, then, parapsychology as an explanation for the twin bond, if it exists, has to rely on the statistical evidence of the correlations of probabilities, rather than more direct evidence. And his evidence is, by its very nature, somewhat unsatisfactory. The Eaves and Last study shows that MZ twins often *think* they have a 'bond' when in fact there is no objective evidence for it.

## Astrology and twins

Twins, by being conceived and born at the same time, are of the same astrological sign, and this is sometimes given as a reason for their close psychological similarity. There are two things to be said about this.

The first is very simple. If astrology were to account for the coincidences, then these should be just as apparent between DZ twins as for identicals. All twins, whatever their zygosity and their sex, are (with very rare exceptions) conceived and born at the same moment. There is no difference between MZ and DZ twins on this point. Yet DZ pairs do not show the kind of extraordinary coincidences seen in MZ twins. Or, if they do, the coincidences are not noticed to anything like the same extent.

It is often said that the differences between MZ and DZ twins lies not so much in their genetic make-up as in the fact that their parents treat MZ twins more alike and, in doing so, *create* similarities that would not otherwise be there. If parents do create similarities in MZ twins raised together, it is then argued, these practices might obscure any astrological similarities that exist between the different types of twin. After all, DZ twins are more alike than brothers and sisters or unrelated children – so is that due to astrology?

The short answer is again: no. There have been several studies which throw light on this. In the first place, observation studies show that the parents of twins who look alike do treat them more equally than when the twins do not resemble each other closely. However, one study also showed that, as the twins get older, this greater similarity in treatment *does not make the twins more similar in personality*. Two other studies looked at the behaviour of mothers who were mistaken about the zygosity of their twins. In some cases mothers thought their twins were DZ when in fact they were MZ, and in some cases the reverse. In both instances it was found that mothers reacted to the similarities between their twins and did not create them. That is, mothers who *thought* their twins were MZ treated them as much more alike than when they *thought* they were DZ. It surely follows from this that parents respond to, rather than create, similarities in MZ twins and so do not distort the picture as between MZ and DZ.

It further follows that the differences between MZ and DZ twins, not being distorted by parental upbringing, must be due to genetic and environmental influences. Astrological influence, if such there is, cannot account for the coincidences between twins.

A brief summary, now, of the main points so far in this chapter. In the first place, I cannot stress too much that rare events do happen. People *do* die in airplane crashes and get struck by lightning – they even make killings on the tables at Monte Carlo and Las Vegas. In 1976, during a party aboard the *Queen Elizabeth II*, on her way from Southampton to New York, a well-bred young Englishman fell overboard. He freely admitted he was blind drunk. His absence was not noticed immediately and it was a full half-hour before the captain could be sure he had fallen into the Atlantic. But the captain turned the ship around, 'on a button', and retraced what was left of his wake. Astounding as it may seem, the Englishman was found thrashing about in the water, his champagne glass still in his hand. The odds against finding someone in an ocean swell like that must be enormous – but it happened.

Next, the coincidences themselves. This chapter shows that many of the similarities in behaviour which the Minnesota twins share are not – when it comes down to it – as rare as you might think. True, we were not able to consider all of them, including some of the apparently more bizarre coincidences, like the two Jims both having a white seat around the tree in their garden, or the tendency for Jack and Oskar both to sneeze loudly to surprise people. But we did consider equivalent coincidences – equivalent in the sense of seeming to be equally rare – and the figures for these, though large, were not fantastically rare, in comparison with other events that *are* known to happen even several times a year.

If both twins have a fear of heights, it *seems* highly unusual. But this similarity between them would, we must remember, occur *simply by chance* once in every two or three sets of twins, depending on sex. Sixteen pairs of twins have been seen at Minnesota, so the laws of chance alone predict that at least three to five pairs of twins would share such fears – not so very different from what Bouchard and his colleagues found.

Add to that the finding that twins do not appear to share any 'psychic bond' such as ESP or a supernatural understanding of each other, and that astrology is no help either, and the startling notion of 'uncanny' coincidence begins to fade.

But I am not saying that nothing abnormal is going on: everything is not yet 'explained away'. Twins do not offer evidence for the validity of astrology or parapsychology, it is true. But there is one result that has emerged from this chapter that we have not yet returned to.

This is the fact that twins do not share just one or two coincidences together, but many: in the case of the Jims at least twenty-one; twenty-seven for Barbara and Daphne; eight that we know of for Keith and Jake; and so on. We have seen that the odds against the combinations of coincidences that we *can* measure are as high as 1 in 50,000,000. Clearly, if we could take into account, say, all twenty-seven coincidences between Barbara and Daphne we would end up with still higher odds. Even so, we should be careful before reading too much into these figures.

In the first place, we simply do not know how many coincidences, on average, any two people of the same age, sex and place of birth picked at random would share. The research has just not been done. We know that 46 per cent of people fear snakes and we know that 59 per cent of people drink coffee at breakfast, but we do not know the proportion of people who show both these characteristics. Still less can we know how many people have twenty-seven characteristics in common.

Many statisticians would say that this is the wrong approach in any case. Suppose you take a pair of twins and examine thirty characteristics; suppose further that there is a 1 in 5 chance that the twins share any given characteristic (and we have seen from this chapter that many of the coincidences at Minnesota in fact occur with roughly this frequency); then you would expect the twins under examination to share about six of the thirty characteristics. Yet if you then ask: what are the odds of these particular six characteristics being shared, the answer is $5^6$ – or 1 in 15,625. Those odds *look* long but they are of course what you would *expect*.

This may be a difficult point to grasp but we need to remember that although one pair of twins may share twenty-seven characteristics, we have to set that against the hundreds, even thousands of characteristics which any person has and which are capable of being shared with someone else. As one statistician put it to me: 'I bet I could take, say, Adolf Hitler and the chief rabbi and find twenty-seven similar characteristics, where the particular odds of them occurring are millions to one.' Just as the world is a big place, with a lot going on, so an individual is a collection of a very large number of characteristics; that is why we are all unique. Scientifically speaking there is a flaw in the way the coincidences at Minnesota are being collected. The only way to be absolutely certain that there are connections between identical twins is to draw up a list of characteristics *before* they arrive in the laboratory and then to see if the number of coincidences between them is higher than between two unrelated people of the same age and sex. Now that Professor Bouchard has seen fifteen or sixteen pairs of twins, perhaps he will remedy the methodological flaw with future visitors to the twin cities.

A second reason why we have to be sceptical about the apparently unusual coincidences observed at Minnesota is that we do not know how many of the characteristics are actually related to each other in some way, psychologically speaking. Many things, like reading preferences, phobias, nailbiting, bedwetting, piano-playing or sneezing for the effect it has on others, are possibly expressions of the personality; might inheriting a particular set of genes which help govern the personality make a particular chain of characteristics far more likely? And therefore would anyone with the same genes show the same set of habits.

For my part, two riders are worth adding to these reservations. Despite the fact that nailbiting, bedwetting, a fear of heights or a love for vodka may all be relatively common activities, so that coincidences between twins along these lines are not really so surprising; and despite the fact that even psychologists suffer from the 'halo' effect (that is, spotting patterns that aren't there, as the numerologists do), none the less *some* of the coincidences at Minnesota are just too unlikely to explain away. These are mainly those striking coincidences where the odds are particularly difficult to calculate: the fact that both Jims built white benches in their back yards, that Bridget and Dorothy filled out the same diary for the same year, that several pairs of twins enjoyed the same authors (even when they wrote under different names) and that at least two sets of twins gave their children the same names. In other words, we have two kinds of coincidence at Minnesota: those which, by themselves, don't mean anything, and those (which we can't measure) which do.

Secondly, even after the statisticians have had their say, it still seems that the odds against whole chains of coincidences occurring are so huge that something about human behaviour is being reflected in the Minnesota findings.

In Susan Farber's book, *Identical Twins Reared Apart*, she analyses almost every aspect of twin life – except the gestures, mannerisms and habits, which are the things that attract everybody's eye. This chapter has looked at these aspects of behaviour and tried to gauge the extent of the 'halo' effect. What we are left with, it seems to me, is something like this: forget astrology, forget 'psychic bond', forget – for the sake of argument – all those shared similarities that, like fears, are not very rare once you can attach a precise figure to this probability. But concentrate instead on the fact that a whole chain of coincidences seems to hang together: does this tell us something about the way personality is organized and how *behaviour* reflects gene action in all sorts of little ways? In other words, are all the coincidences that are being collected at Minnesota a sort of camouflage, a signal for something else that is going on at a deeper level?

# 6  Coincidences around the World

The coincidences at Minnesota have attracted attention mainly because of their very specific nature: from shirt styles and jewellery to children's names and funny habits like 'squidging' up noses. Yet the whole point to twin research, at all times and wherever it is carried on, is the search for coincidence. Psychologists, psychiatrists and paediatricians do not put it like that: they search instead for 'concordance' and 'discordance' between twins. If they want to know whether, say, cancer is inherited they look at the proportion of twins who are 'concordant for this characteristic' – that is, twins both of whom have cancer or both of whom do not. They speak of twins who are 'concordant for' heart disease, 'concordant for' schizophrenia, 'discordant for' drug abuse or 'discordant for' measles. Essentially, they mean the same as you or I when we use the word coincidence. The vocabulary is slightly different but the meaning is much the same.

Put like this we can see that the Minnesota study is essentially no different from any other study of twins. Thomas Bouchard's unit is just one of a large but growing number of studies looking at twins and the 'concordance' between them.

This chapter places the Minneapolis research in proper perspective. At the moment, twin studies in general are undergoing a resurgence of interest. Changing attitudes to illegitimacy mean that very few twins in the future will be adopted into different families, so this is likely to be the last chance we will ever have to study twins reared apart. The past two decades have seen the social sciences dominated by psychologists and psychiatrists who have studied the effects of environment on behaviour, yet have not made as much headway as they anticipated in understanding such things as personality, crime or drug abuse. A new interest in 'behaviour genetics' has grown up, partly as a result of this disappointment, and twin studies are an aspect of that.

The research we shall be covering is interesting enough in itself but a lot of it, especially that later in the chapter, should help us to answer this question: how much more alike, *biologically and psychologically*, are twins compared with two average people who bump into each other on the street? If we can say, for instance, that twins are ten times more alike than unrelated people, on a particular dimension, then coincidences between them in associated areas should be ten times more likely and, on the same grounds, ten times *less* surprising. If for instance, twins are ten times as alike as unrelated people, psychologically speaking, then the fact that they share the same phobias which occur in 46 per cent of the population should come as no surprise at all. Knowing precisely how close twins are compared with the rest of us may enable us to modify the calculations we made in the last chapter. In doing so, we may change our understanding of the coincidences revealed at Minneapolis.

Twin research has had a bad press lately, especially after the frauds of Sir Cyril Burt. Nevertheless in 1973, at the very time that Sir Cyril's frauds were being exposed, the International Society for the Study of Twins was being formed. The Society has so far held three international meetings, the latest in June 1980 in Jerusalem, only shortly after the first news of the extraordinary coincidences at Minneapolis. The conference was held at the Gedda Institute, which was itself opened on 20 June of that year, the day the conference closed. Like its counterpart in Rome (also founded by Luigi Gedda) the institute provides free medical advice for twins, in return for which the twins make themselves available for research. The Jerusalem conference attracted 149 papers from eighteen countries, including – besides western Europe and North America – Japan, Israel, Nigeria, Singapore and Australia. Much of what follows draws upon papers given at the Jerusalem Conference.

## The twin registers

The nature of twinning has given rise to a particular methodology, the twin register. In various places around the world doctors, psychologists and sociologists have set up twin registers so as to be better able to study twins.

These vary in size from a few hundred to several thousand. In some cases only young children are chosen to start with, and followed through into adult life; in other cases only ex-soldiers are included. The first began in Sweden in 1896; the latest, started in Australia in 1976, already includes 11,000 pairs of twins. The best-known twin registers are:

*The Louisville Twin Study, Kentucky.* This study has been publishing its results since 1967. In the main these have involved the detailed follow-up of a relatively small number of twin pairs, very often throughout several years of their lives. Examples of Louisville studies include interviews with the mothers of 140 pre-school twins to check links between birth weight, birth sequence and behavioural differences (like sleep disorders). Other studies have looked at the development of twins' grammar, feeding problems, intelligence, activity levels, accidents, personality development, speech and reading; in other words, fairly straightforward paediatric data about growing children who happen to be twins. One major finding of the Louisville study concerns the mental development of twins. The Louisville study has found that, as they grow older, monozygotic twins become more concordant and also parallel each other for spurts and lags in development. Dizygotic twins, on the other hand, do not show this: in fact they grow less concordant and show fewer coincidences with age.

*The Berlin Twin Study, Institute for Human Genetics and Anthropology, Berlin.* Compared with other twin *registers* this, too, is a fairly small study, but it has followed through its twins for forty years. The study, started during the Second World War, covers 265 pairs of monozygotic twins and 230 pairs of DZs, in both cases divided almost equally as to males and females. The first follow-up of the Berlin twins was carried out in 1954–5 and has been repeated regularly ever since. Interestingly, the study also includes three sets of identical triplets, three sets of dizygotic triplets and one of tetrazygotic quadruplets. Many of the Berlin twins have already died so the study also offers a good opportunity to study ageing, death and the illnesses of twins. A major thrust of the research by Dr Gerhard Koch, director of the Berlin study, has been in the field of cancer; his main finding is that genetics does not appear to be a major factor in cancer.

*The National Academy of Sciences – National Research Council Twin Registry.* This is a *register* not a study as such. It consists of a panel of 15,000 twin pairs, all born between 1917 and 1927 in the United States, all male, and all members of the armed forces (medical records generally being much better for soldiers than for others). In addition, a further 15,000 twins on the panel are or were in the service, but not enough is known about their twin brother.

Academics are allowed access to these twins for research purposes, provided certain criteria are met (for example, that the answers they want can only be got through twin studies). But the NAS–NRC also sends out a questionnaire to obtain basic information on these twins such as vital statistics, military, personnel and criminal records, medical records and even fingerprint records. The panel has so far provided useful results about the aetiology of cardiovascular disease, various types of psychopathology, headache, the link between pollution and respiratory disease, and factors affecting socioeconomic success. In the future the study will probably turn to the genetic problems surrounding ageing and death. Preliminary data is already being collected about this.

*The European Twin Registry, Division of Human Genetics, University of Leuven, Belgium.* This new registry is a collaborative effort among forty teaching hospitals within the European Community. A major interest of the doctors running this registry is the assessment of prenatal and perinatal care at multiple births. The registry uses teaching hospitals as only they are equipped to carry out postmortems on babies who die very young. The hospitals pool their resources on suitable births, collecting blood and placental specimens to determine if the children are identical and providing information on medical care, delivery procedures, malformations and mortalities. The aim is to provide *conclusive* data on which malformations are more common in twins.

The European registry is also interested in the differences *between* monozygotic twins according to when the division of the cells takes place. It is now known that 32 per cent of MZ twins divide in the first three days after conception, 64 per cent between day four and day nine, and 4 per cent from the tenth day on. Congenital abnormalities may be related to the time of cell division, although this is not as yet certain. (It is thought, for instance, that those rare cases where one twin is born without a heart may be associated with late division – after ten days.) The other areas the registry hopes to study are diseases of the central nervous system (anencephaly, spina bifida, mental retardation) and of the extremities (especially the hip), schizophrenia and multiple sclerosis. The idea that the presence of a healthy twin sometimes contributes to the survival of a malformed co-twin who would otherwise die is also an area of interest in the European registry. The aim is for 10,000 MZ pairs, a figure which, the directors feel, would be enough to study most common abnormalities properly.

*The Finnish Twin Registry, Department of Public Health Science, University of Helsinki, Helsinki.* The Finnish registry was established in 1974. Its first cohort consisted of adult like-sexed twin pairs, both of whose members were alive in 1967. The twin population was picked from the computerized files of the Central Population Registry, which includes all Finnish citizens, including those living abroad and those dependent on institutional care. There are about 16,000 twin pairs in the registry, plus another 1429 with one or both members dead in 1974 but for whom data exists. So far the registry has mostly studied medical factors – angina, pain, chronic cough, a history of breathlessness, weight and dieting, drugs and contraceptives. Other behaviour, however, of relevance to the Minneapolis study has also been investigated – drinking habits, sleep, pets – as well as details about fears and general satisfaction with life. The Finnish study, being based on such an excellent sample, provides very good fundamental figures on twins.

*The Norwegian Twin Register, Institute of Medical Genetics, Oslo.* Three major studies have been carried out using Norway's traditional record of all births, the Central Bureau of Statistics. In one study in 1965, all twins born between 1901 and 1930 – 25,000 pairs – were checked against the national register of psychoses. In 342 of these pairs, one or both twins had been hospitalized for schizophrenia, manic-depression or reactive psychosis. The analysis showed that genetics *was* a factor in schizophrenia but a weaker factor than was previously thought. Other studies of the Norwegian registry have related to criminality, neurosis and coronary heart disease.

*The Australian Twin Registry, Australian National University, Canberra.* Attempts are currently being made to register every pair of twins in Australia. There should be 100,000 pairs but the registry will happily settle for half that. At the time of writing they have 11,000 pairs from all over the continent. Like most registries all *bona fide* researchers can make use of the panel of twins provided the research proposals pass an advisory committee. The first study to use the register is already in progress in Sydney, where twins are being compared for the effects of alcoholic intoxication on driving skills. Other projects already planned include the causes of coronary heart disease, hyperactivity in children and the study of personality in adult twins.

There are several other twin registries – in London (which has the only psychiatric twin register, based on admissions to *one* hospital), in Copenhagen, in San Francisco – but those mentioned above comprise the biggest and most comprehensive in terms of access to the twins of a country.

## Why twins are getting scarcer

DZ twins, but not MZs, are getting scarcer in many parts of the western world. Dr Ian MacGillivray, of the University of Aberdeen, has studied those factors which may be associated with the decline of twinning in Scotland. He has found that the percentage of mothers who had multiple births in Scotland remained stable from 1856, the first year for which he looked at the records, to 1957, when it began to fall.

There are – surprisingly perhaps – no known factors which affect the rate of MZ twinning but many things are known to affect DZs. MacGillivray considered height, weight for height, social class, nutrition, cigarette smoking, age, parity and fertility. In Aberdeen at least, height is important in twinning – tall women are more likely to have twins than shorter ones. The same is true of heavier women. However, neither of these factors has changed appreciably in Scotland, so they cannot account for the smaller number of twins. MacGillivray ascribes the decline to the fact that, owing to contraception, women are having fewer children and having them younger. The two are linked: women are having their first child no earlier than they ever did, but they are having smaller families over a shorter time, so the average number of children and the average age of mothers at childbirth are both going down. It is known that twinning is associated with children who are born later and mothers who are older.

It is also just possible that the contraceptive pill reduces activity of the pituitary gland and that this in turn prevents multiple ovulation. Probably both sets of factors account for the decline in DZ twins.

A study by doctors in Oxford and Aberdeen found the mothers of DZ twins to be both taller than women in the general population and in comparison with the mothers of MZ twins. Since there is some evidence that body size is governed by the pituitary gland, this is indirect evidence in support of the pituitary theory.

More fertile mothers are also more likely to have twins and this, too, may account for the decline in twinning. DZ twinning is known to be associated with illegitimate or premarital pregnancies: fertile women, more easily 'caught' by an unwanted pregnancy, are more likely to have twins as well. Children conceived in the first three months of marriage are also more likely to be twins. It was found in Italy that when men were discharged from the army after the Second World War, the average time to their wives' conception was 2.2 months less in the case of twins. It is perhaps only natural, now that contraceptives are widely available, for fertile women to make most use of them. A reduction in the twinning rate would follow.

Some scientists, however, think the drop too rapid to be accounted for simply in these terms: they say that the male sperm count must have dropped also, perhaps due to the increased use of pesticides in industrialized countries. Still other scientists suspect the role of anabolic steroids in foodstuffs. No one can say with certainty.

In Italy, where the records are unusually good (the Catholic Church is mainly responsible), it is now known that there was a peak of twinning after the First World War, followed by a high plateau between the wars and a decline since 1945, which has been especially sharp in recent years. If we accept that Italians used contraceptives much less than other west Europeans, then the fall in Italian twinning cannot be put down to the same causes as may have applied, for example, in Scotland. If the women were not using contraceptives they were presumably continuing to have children into the 'twin-prone' years. Moreover, records show that, in the 1930s, when twinning rates were high in Italy, the age of mothers, and the number of children they gave birth to, was already falling. The decline in twinning started during the Second World War, when maternal ages rose for obvious reasons. This, too, goes against the expected pattern: older mothers are supposed to mean *more* twins. Finally, the twin rate (and the birth rate) in Italy has declined overall over the last 100 years, so the pesticide argument also begins to lose force, pesticides being a much more recent phenomenon. The Italian doctors who addressed the Jerusalem conference ended without making any firm conclusions for the trends in their country beyond saying that their figures may reveal a more fundamental change in the world's fertility rates, the reasons for which are still obscure.

Race is known to have an effect on twinning. A US study from Virginia, presented to the conference in Israel, was unusual in providing an opportunity to compare black and white twin rates. The Virginia data, which show an overall decrease in twinning and a marked decrease during 1930–31 and 1939–41, offer another explanation for the change we have been discussing: that during times of depression and war people are stressed and sexual activity is reduced in such a way as to reduce the probability of double fertilization. This explanation, too, is plausible but very circumstantial. The Virginia scientists however offer it as one possible reason why twinning is decreasing most in modern industrial societies, where stress and anxiety are greatest: according to other papers given at Jerusalem, the twinning rate is falling in Japan and France whereas in Nigeria it is still very high – 50 per 1000 maternities. It also seems that the greater the age difference between the mother and father, the greater the chances of having twins. This is a new result and not, as yet, confirmed.

## Mother or father to blame?

It has been accepted for some time that only the tendency to have DZ twins is inherited. Luigi Gedda, in Rome, has produced some evidence to suggest that MZ twinning can run in families but, so far as I know, no one else has reproduced his findings. It has also been accepted that the inheritance of DZ twins is mainly, perhaps solely, carried through the maternal line. However, an Italian study suggesting that paternal inheritance may also be possible was reported at Jerusalem.

The Italian doctors studied 14,725 maternities. Going back two generations and examining the birth records of mothers, grandmothers, fathers and grandfathers, brothers and sisters, they found that DZ twinning showed through on the mother's side *and* that it went back on the father's side too. It seemed, moreover, that different-sexed DZ twins were only found in the mother's line whereas same-sexed DZ twins were found only in the father's line. The Italians do not regard the second result as firm yet, but it does seem that the old wives' tale about twins being inherited through the mother may not be true. (Another complicating factor is the recent discovery that, even in Catholic countries like Italy, the abortion rate is surprisingly high. Abortions involve conceptions, of course, and since in many cases the sex of aborted children, and whether they are multiple foetuses, is not reported or recorded, it follows that the proportion of sons, daughters and twins in families may be distorted by this omission. Possibly, this distortion is substantial.)

## How to have twins

Orthodox Jewish groups observe ritual sexual separation each month during menstruation. Intriguingly, this has thrown light on how to have twins.

Doctors have found that among those Jews who resume intercourse before ovulation, the rate of multiple births is 10.5 per 1000 deliveries; but it ranges from 21.1 to 36.4 per 1000 in those resuming intercourse on the day of ovulation or within the three or four days which follow it. Jews, however, are a genetically distinct group, with certain genetic disorders, like Tay-Sachs disease, which is very common among their children but almost unknown in other groups, so this result may not apply to non-Jews. The timing of intercourse has also been implicated in some hereditary disorders – such as Down's syndrome (mongolism). Some doctors believe that if the sperm have been around for some days before the egg is released, malformation of the foetus is more likely.

Another way to have twins may be to catch a cold. A complicated statistical analysis of twinning in England and Wales, by Dr Philip Burch of the University of Leeds, showed that mothers of DZ twins have a lower age for menopause. Among other things, this suggested to Dr Burch that some women are genetically prone to have twins, but he then found that twinning varied from year to year – a rhythm that seemed to need some other explanation. Looking at this rhythm, he found that the pattern was very similar to that encountered in infectious and allergic disorders. This gave him the idea that twinning might be a by-product of an infection or an allergy, where the infectious or allergic micro-organism might compete for antibodies in the blood that would otherwise suppress a naturally produced clone within the body. In support of his idea is the fact that there is a seasonality in twinning: it is highest in births in November (conception in February) and lowest in May (conception in September). The pattern is repeated in cattle (who have a lot of twins) but Burch is continuing this intriguing line of inquiry.

## How to test whether you are having twins

The techniques of ultrasound are now sophisticated enough not merely to tell you if you are having twins but whether or not they are identical. One report in Israel described how scientists can check the number of membranes surrounding the foetuses: one membrane indicating MZ, two DZ. This technique can also tell us whether there are any problems on the horizon, like the transfusion syndrome, when blood flows from a 'donor' twin into its partner, leaving the donor starved. Swedish doctors are also now able to measure the blood flow in the foetuses and can tell from this whether it is being pumped faster in one twin than the other. This would indicate which is the 'donor' twin. (On the other hand, another Swedish doctor has found that a simple tape measure round the abdomen is very nearly as good as ultrasound in the detection of twins!)

## Are twins the same age?

Of course, one twin *has* to be born first, and is so many minutes 'older' than the other. This tiny difference apart, in many cases the two halves of a twin pair are born at different stages of development – and this difference may be important later in life.

These differences may be partly due to the imprecise nature of the techniques doctors use to assess the developmental age of twins. Since twins tend to be smaller than singletons, the usual tests of development are less reliable for them. Yet one study from the Louisville team found that 34 per cent of seventy-six sets of twins showed discrepancies in their developmental age of between two and six and a half weeks. Some kind of difference was found in as many as 70 per cent of twins. The results were based on a physical examination, neurological test, a clinical assessment and ultrasound cephalometry. Even if a large part of this result is due to error in the techniques, some of the differences must have been due to the twins developing at different rates. We saw in earlier chapters how it is possible for twins to compete within the womb and how MZ twins are often very different in appearance at birth. One twin may be quite normal, in the gestational sense, whereas the other may be premature. And premature birth is statistically linked to later disorders: low IQ, a higher incidence of deviant behaviour patterns, poor academic performance, and so on.

## The armadillo factor – or why you should prolong pregnancy and how to do it

Early detection of twin pregnancy is important, so that precautions can be taken to avoid the tendency for twins to be born prematurely. One common precaution is bed rest, and this undoubtedly prolongs pregnancy and reduces complications. A team of doctors from Singapore has tried a different approach. Two or more foetuses in the mother's womb means two or three times the number of glands, producing more than the usual amount of hormones. These may find their way into the mother's bloodstream, alter her metabolism, and induce abnormalities in the foetuses. The Singapore doctors found that the levels of certain important hormones *are* higher in the mothers of twins and triplets than in mothers with a single child. And they have found that administering drugs to correct this imbalance is more effective than bed rest alone in prolonging pregnancy.

Some doctors are also interested in the armadillo, for the light that animal may throw on human twinning. Armadillos can have up to twelve 'identical' MZ offspring yet rarely, if ever, do they have anything like the problems, or birth abnormalities, that are associated with human twinning. Although the basic genetics of DZ twins are well understood, there is still some way to go in understanding MZ twins in animal species as well as in humans. For instance, it has only recently been discovered that male tortoiseshell cats are, in fact, twins, each *one* of them. They begin life as *two* animals in the womb, two unlike-sexed twins, in fact. Then they fuse and are born as one animal though they actually carry all the genetic material of *both* the original twins.

Similar strange things happen in humans too. In some cases of twin births, one twin has no heart. This has confused doctors for some time, but now Dr Nance, from the Virginia team mentioned earlier, has evidence that the following happens: first one sperm fertilizes one egg; then, as this egg divides, another sperm fertilizes a polar body (that is, a cell without a nucleus) and not a second egg. This is an important line of research, since some of the Australian doctors also believe that polar bodies are in some way associated with fertility drugs. Maybe the drugs indirectly cause the birth of twins with no heart?

## Genes, Hiroshima and falling downstairs

Twin research over the last few years has seen a decline in studies of the physical characteristics of twins – height, weight, pulse rate and so on – and an increased interest in their behaviour, life-styles and diseases. This is where the most important work lies and, indeed, where the biggest surprises are in store. For instance, research on Mormon twins, where there are complete family records going back to 1800, and on twins in Japan, where one twin was exposed to the atom bomb explosion and the other was not, all show that identical twins are more likely to die at the same age. But – and this is extraordinary – the pattern applies equally to death from old age or disease and to death from accidents. Scientists are asking: is there a genetic basis to accident-proneness? (Remember Barbara Harrison and Daphne Goodship both fell downstairs at the same time.)

It is now time to consider the psychology of twins in some detail. All of the topics which follow – alcoholism, heart disease, crime – have a psychological side to their aetiology. They may also have a genetic aspect, too. No one doubts that psychology is involved: it is the idea that genes may play a part in crime, low IQ or drug abuse that sparks excitement and curiosity.

## Drink

The tendency to drink is inherited very weakly, if at all. The Finnish twin register has looked at this and found that the drinking habits of MZ twins were more similar than those of DZ twins, but the differences were not very large and varied greatly from age to age (a study of 600 twin pairs in Britain found some evidence for genetic influence but that it differed as between men and women); in the older age groups what differences there were disappeared, suggesting that the earlier similarities may at least in part have been due to familial factors which applied less later on in life; the effects, though weak, were much stronger in men than in women.

On the other hand, the tendency for heavy drinking appears to be more strongly inherited, at least in so far as the concordance within MZ pairs is much stronger than in DZs. It is also stronger in women than men.

When we come to alcoholism proper the picture is somewhat different. Quite a lot of research has now gone into this subject and, following the Jerusalem conference, fairly firm conclusions can be drawn. Alcoholism seems to run in families. Roughly speaking, a third of alcoholics have one or more alcoholic parents – the same is true of only 5 per cent of the general population. It has also proved possible to breed strains of animals (usually rats or mice) that prefer weak alcohol solutions to water or else that avoid alcohol at all costs. What seems to happen is this: people who have alcohol in their family background do not differ from others in their rate of metabolism of alcohol but they do develop significantly higher levels of an intermediary metabolite, acetaldehyde. This may affect the level of tolerance someone has to the effects of alcohol or their propensity to physical dependence. It is certainly the case, for example, that young men with family histories of alcoholism relax more (in their muscles) while their blood alcohol is rising and experience less intoxication than others. So maybe what they inherit is a high level of tolerance.

Twin studies comparing MZs and DZs have confirmed a high level of heritability for alcohol elimination and for the manufacture of acetaldehyde; if alcoholism is inherited to any degree, this may well be how it happens. Only one study has tried to assess directly the 'concordance' rate; this study looked at forty-five pairs of twins, and used a fairly rigorous definition of alcoholism. The doctors found that the concordance rate for MZ twins was 58 per cent, for DZs 28 per cent.*

Marc Shuckitt, a psychiatrist working in the Alcohol Treatment Program at the Veterans Administration Center in San Diego, California, believes that the level of acetaldehyde may eventually be a useful marker in identifying people who are biologically prone to suffer alcoholism. However, no one has yet worked out how this marker would be used. It is by no means certain that simply telling someone he is prone to alcoholism would deter him or her. This study has not been replicated.

Twin studies reported at Jerusalem prove – if proof were still needed – that drink in too large a quantity actually destroys brain cells. A British comparison of the brains of twenty-five MZ twin pairs, where one drank and the other did not, shows this. (The study also showed, incidentally, *no* genetic component for alcoholism – such are the contradictions of twin research.)

The effect of genes on drinking behaviour is by no means straightforward: research in Sydney, Australia, shows that even twins vary enormously in their behavioural reactions to drink. The psychologists (who were being paid by the brewers) were interested in the effects of drink on driving, so they tested twins for vision, hand–eye coordination, judgement of speed and so on, before and after taking alcohol. They did not find that MZ twins were any more similar in the change that came over them than DZs. What they did find was that women appear to suffer less than men. But, as they themselves conceded at Jerusalem, their study was still in its early stages (seventy-nine twin pairs only studied to date).

It seems that alcohol does have a role in angina pectoris (chest pains). Zdenek Hrubec examined the records of 1200 male twin pairs from the registry at the Karolinska Institute

---

* A 58 per cent 'concordance rate' can mean one of two things: *either* that 58 per cent of a sample of twins *both* share a characteristic; *or* that, once you have a sample of twins, all of whom show that characteristic, 58 per cent of their co-twins share it. These are not quite the same things, but in both cases the genetic influence of a characteristic is assessed by the difference in concordance rates between MZs and DZs. To recap, all the differences between MZs *must* be due to environmental influence, whereas differences for DZs may be due either to genetics or to environment. It follows that the *difference* between MZ and DZ concordance rates is a measure of the genetic influence.

in Stockholm, and 4000 male twin pairs from the NRC twin registry in Washington. He looked at the drinking patterns of the men, their exercise habits and their career progress. He found that of these three factors only alcohol consumption was related to angina in the MZs. It was known before that angina was related to smoking but Hrubec's study was the first to note the connection with drink. He also found a strong connection between drinking and smoking in MZ twins (as well as in the general population, it should be said). At the end of his paper he raises the possibility that the link between smoking and drinking may be strong enough to influence the apparent link between smoking and angina: that is, angina may be due as much to drinking as to smoking, if not more.

## Coffee and tea

Strange as it may seem, coffee-drinking appears to have a hereditary component whereas tea-drinking does not. The Finnish twin registry showed that MZ twins are half as likely again as DZ twins to be concordant for coffee-drinking. However, as with alcohol, the difference between MZ and DZ twins decreases with advancing age. By itself this could mean that younger twins, living in the same house before they are married or independent, influence each other to drink coffee. Then, as they grow older, they move apart and lead separate lives. If this is so, the pattern revealed by the figure is not necessarily evidence of a genetic component. But why is it not repeated for tea-drinking and, for that matter, sugar usage, which like tea also appears to lack a hereditary component? The Finnish registry shows no consistent pattern of difference between MZs and DZs on tea-drinking or the number of sugar lumps used. This makes the finding on coffee all the more interesting. (Remember that in the Minnesota study, the twins listed similar alcohol- and coffee-drinking patterns, but no one had anything unusual to report about tea or sugar.)

## Drugs

The uses of many drugs are determined by the diseases that make them necessary, and when the illness is known or thought to be hereditary (e.g. heart disease) not much is gained by looking at drug use. For other drugs, however, the picture is somewhat different. The Finnish study compared how much more likely twins were to share drug habits than any two, equally similar, non-twins. Their results may be summarized like this:

|  | *MZ* | *DZ* |
|---|---|---|
| Anti-hypertension drugs | 28 | 11 (up to age 40, dropping off after) |
| Hypnotic drugs | 76 | 1 (i.e. exactly the same as for unrelated people) |
| Tranquillizers | 13–14 | Not available |

Even with these drugs, however, their use may well be associated in some way with 'nerves' or some other underlying personality problem which could be genetic. Only drugs like amphetamine or cannabis are obviously unrelated to this. The use of dextro-amphetamine and people's responses to it have been examined by John Nurnberger and his colleagues at the Biological Psychiatry Branch of the National Institutes of Mental Health just outside Washington. They studied thirteen pairs of MZ twins aged eighteen to thirty-nine. They gave the twins either a placebo or dextro-amphetamine in random order and on separate days, and studied their responses – motor activity, amphetamine plasma level – and the answer to a standard mood questionnaire, which was completed during the hour after the drug was given. The questionnaire covered feelings of depression, crying, increased or decreased speech, sexual preoccupation, anger/irritability, delusions, sleepiness and laughter.

Nurnberger found that three factors were highly concordant for the twins. The strongest response occurred thirty to sixty minutes after the drug was given and they called this 'excitation' (the concordance ratio being 79 per cent). This factor included increased speech, demanding contact, distractibility, speech disorganization and anger. (It did not, however, correlate significantly with amphetamine plasma level.)

A second factor, which they called 'elation', also emerged.

This too occurred during the thirty to sixty minute period after the drug was given and consisted of laughter, increased activity, and an absence of anger. This factor *was* correlated with amphetamine level in the blood. A third factor – depression within the first twenty minutes – was also found, but not very strongly.

Nurnberger also tested the brainwaves of the twins and found a high correlation between the way the drug changed the waves of one twin and the way it changed the waves of the other. This was true of both sides of the brain. There were, finally, concordant changes in some (but not all) hormone levels in these twins.

Nurnberger believes that these results taken together imply individual (genetic) differences in substances in the brain known as neurotransmitters, which are affected by amphetamines. This is important because it could mean that some people are much less responsive to amphetamine than others and are therefore less likely to abuse it. That could matter to a doctor on the lookout for what (and what not) to prescribe. This result could also explain the observed sex difference in relation to amphetamines (women are more often depressed than men).

## Contraceptive pills

Both MZ twins and DZs were more likely to be concordant in their use of the pill than pure chance would suggest. The figures were not high – only four or five times higher than chance for MZs, and twice as high for DZs. Still, they are worth a thought. MZ twins look alike; we can expect this to affect their sexual experience (all other things being equal, good-looking people have more sexual success). In turn this affects the need to use contraceptives. A genetic characteristic, such as looks, may influence a social one, such as use of the pill. This is interesting if we compare the figures on the pill with those for drugs in the last section.

Female MZ twins were up to seventy-six times more concordant on some medical drugs (the hypnotic ones) whereas on the pill they were only five times more likely than any two women of their age to share the habit. These figures may be a useful guide to the relative genetic components of medical and social characteristics that we all share. In other words, contraceptive pill use (with a value of five) has a much weaker genetic component than hypnotic drug use (seventy-six).

## Heart disease

Heredity and heart disease is an extremely complicated issue. There seem to be some inherited characteristics which predispose a person to heart disease, but no one can agree on exactly what they are or on how important each one is. Quite large studies, by well-respected doctors, contradict each other. At the risk of oversimplification I discuss four aspects: blood pressure; blood content (sugar, fats); stress; smoking.

### Blood pressure

Blood pressure seems to be mostly determined by heredity, at least according to researchers at the National Research Council Twin Panel. On the other hand, studies at the University of Indiana on the families of twins show that the inheritance of blood pressure itself has been overrated; what actually happens is that our genes control body size, which in turn affects blood pressure. This study particularly showed that the connection between a mother's blood pressure and that of her offspring is much weaker than had been thought. It also appears that people tend – for some reason – to marry people who have similar blood pressure. So if spouses are not taken into account the influence of any single parent can be exaggerated.

### Blood content

The rate at which sugar is drained from the blood does appear to be determined by heredity, and a high level of sugar, or a slow rate of draining, are risk factors for heart trouble.

The disagreements here start over the fats in the blood. The NRC study concluded that heredity did not seem to matter in regard to blood cholesterol but that it may be important in other kinds of blood fats. The NRC researchers found this surprising since, in a few families, cholesterol level is known to be highly heritable. However, Kare Berg, Professor of Medicine at the University of Oslo, has just completed a study of 5500 twins and found a different picture. He noticed that the MZ and DZ twins in his study had different patterns of dying. The twins had all been born between 1915 and 1945 but whereas 58 per cent of the twins who were both still alive were DZs, only 42 per cent were MZs. Compare this with those twin pairs where one had died and the other was still alive: 65 per cent were DZs and 35 per cent MZ. In other words, MZ twins tend to die closer to one another.

This pattern was more marked in the older twins, the ones more likely to have suffered a heart illness. Closer inspection of coronary heart disease in the twins produced this table:

|  | Number of twin pairs | |
|---|---|---|
|  | MZ | DZ |
| Both members affected | 19 | 3 |
| One member affected | 10 | 9 |

These figures are very suggestive of a hereditary component in heart disease. And they are reinforced if you look further into those cases where only one twin had coronary disease: five of the ten MZs had hypertension, but only one of the nine DZs.

Berg therefore looked next at blood constituents. First, he found that cholesterol level had a strong hereditary component, the concordance rate for MZs and DZs being, respectively, 52 per cent and 35 per cent. He found a similar genetic component with lipoproteins but not with triglycerides.

For the moment, therefore, there is evidence for quite a lot of genetic influence on the contents of the blood which predispose someone to heart disease.

*Stress*

Twin research does not always produce only evidence for genetic influences. Einar Kringlen, also from the University of Oslo, has produced a paper, admittedly based on a very small sample, the aim of which is to show the importance of personality and environment on heart disease. For this he contacted seventy-six MZ and DZ twin pairs where one or both had been treated in hospital for coronary heart disease. He found the concordance rates for MZs was only 36 per cent compared with 24 per cent for DZs, and that far and away the most discriminating factor was a 'stressing work situation'. Kringlen gave all his sample a series of questionnaires which asked, among other things, whether they were dominant personalities, whether they smoked, took tranquillizers, drank, commuted to work, did a lot of overtime, including whether they had to work under pressure.

Kringlen reasoned that, since the MZ twins had the same genes, if one had been in hospital for heart trouble and the other had not, then this difference must have been caused by other – non-genetic – factors. He found that the heart patients were more likely to be classified as 'Type A' personalities. This is an American distinction (between Type A and Type B) in which the Type A person is competitive, easily provoked, ambitious, impatient, etc., and the Type B person has the opposite characteristics. Is Kringlen's interpretation watertight? Might not someone whose constitution makes them impatient, ambitious, etc., actually seek out jobs that enable him or her to live in this way? The job therefore becomes an expression of that person's personality or character, as partly determined by heredity, rather than a stressful situation which has accidentally arisen.

Dr Kringlen's study also throws some light on the fourth aspect of heart disease: its relation to smoking.

*Smoking*

Kringlen found smoking to be associated with heart disease, but again his study may only show that those twins with a Type A personality smoked more, which may have had as much to do with their personality as with their jobs – in other words, with their genes as much as the environment.

An NRC study of 514 twins in Massachusetts and California found that in twin pairs where one twin smoked and the other did not, the smokers were more subject to coughs and bronchitis than the nonsmokers. This was true for both identical and fraternal twins. In the fraternal twins it was found that the smokers also had more chest pains (angina) than the nonsmokers but *no* difference existed between identical twins. This probably means that smoking is more important in causing bronchitis than in causing angina. As we have already seen, smoking, once strongly implicated in coronary disease, is now much less so. There can, however, be no doubt about its importance in bronchitis. The NRC study of 11,000 twins showed that smoking more than twenty cigarettes a day greatly increased the rates of bronchitis and prolonged coughing and that smoking plays a bigger part in causing respiratory problems than air pollution, living in towns, or alcohol-drinking (drinking came second).

None of this is particularly surprising. For more thought-provoking results we can turn to a study by Dr G. D. Friedman of the Kaiser-Permanente Medical Care Program at the University of California at Berkeley. He looked at thirty-three pairs of MZ twins discordant for smoking. Friedman found that the smoking twin in a discordant pair started smoking later than the average smoker, tended to smoke fewer cigarettes with less tar content (but with no less inhalation or use of unfiltered cigarettes). Smoking twins also used coffee, alcohol and marijuana more, and on average were thinner. Dr Friedman concludes that smokers may not therefore be matched with nonsmokers on the classic coronary factors (of body-build and substance use) and that twin studies could thus distort the role of genetics in coronary heart disease. We should therefore treat other twin smoking studies with some scepticism. This may apply particularly to the Scandinavian Twin Register studies: the Finnish study found smoking to be a heritable trait, though only to a modest degree, whereas the Swedish Register found coronary heart disease, cough and chest pain just as much in non-smoking twins as in smokers.

The Minnesota workers, despite early results to the contrary, *did* find that lung function differed between smoking and nonsmoking twins reared apart. What they do not say, but was very clear to me, was that the smoking twin always looked older than the nonsmoker: the hair was coarser, the skin was rougher, the general appearance much older. The perfect ad, in fact, for not smoking.

The conclusion to be drawn, from twin studies at least, is that smoking is not quite as bad in its effects as it has sometimes been painted. This is true in part because many smokers are also drinkers – and drink seems to do more damage. It is also the case that smoking can be as much a symptom of an underlying personality pattern as it is a cause of disease. Certainly the thrust of twin research should cause us to change our drinking habits as much as, if not more than, our smoking patterns.

## Cancer

Most cancers do not appear to be simply inherited. Dr J. Kapino and his colleagues from the Finnish Department of Public Health presented evidence at Jerusalem based on a comparison of the Finnish Twin Registry of all same-sex twins born before 1958 and the Finnish Cancer Registry. The prevalence of all types of cancer among all types of twin was much lower than in the general population, and this was especially true in the over sixties. Even among the brothers and sisters of twins who already had cancer the expected ratio was not especially high. In fact, in the entire registry only one fully concordant pair was found, both with prostate cancers. They conclude that 'genetic factors probably have little influence on overall cancer morbidity'.

But wait. Professor Gerhard Koch, the director of the Berlin longitudinal study of twins, found thirty-eight of his 500 twin pairs with carcinomas (thirteen MZ and twenty-five DZ), four with sarcomas (three MZ, one DZ), three MZ pairs with brain tumours, and one with leukaemia. Professor Koch found that, among his twins, breast cancer in women and stomach cancer in men were the most frequent. Although his figures were small he found that the proportions were the same as in the general (German) population, and so this suggested that any genetic influence was weak.

Koch does add, however, that 'although twin studies alone have shown that genetic factors are not of major importance as a cause of common cancers, the question of a genetic precondition is not yet clear. Cancers are different in their aetiology and are genetically heterogeneous.' Koch therefore looked at the *families* of his twins. Here he found that 13 per cent of his thirty-eight twins with cancer had two or more first-degree relatives with cancer somewhere in their bodies. It seems that there is what you might call a 'cancer family syndrome'. In one case, for instance, one MZ female twin was operated on, aged fifty-one, for cancer of the colon. Her twin sister was then examined (one lived in West Germany, the other in West Berlin) and found to have a malignant colon adenoma. The twins' father was also operated on at the age of seventy-four for a colon cancer and four other family members suffered from *different* cancers. Something of a hereditary character seems involved here, but it is by no means clear just how the effect is produced and transmitted.

The Berlin study found that uterine myoma was the most common of the benign tumours (this was the first time a twin study had looked at this problem). Koch found twenty MZ and twenty-one DZ twin pairs with the same myomas, but the concordance in the MZs – 50 per cent – was much higher than in the DZs (9.5 per cent). Koch concluded that genetics plays a much bigger part in the growth of these benign tumours. He also found evidence to support a 'uterine myoma family syndrome', as well as for adenomas of the prostate.

The most thorough study of breast cancer appears to have been completed recently in Denmark. From the twins in the Danish Register, Dr Niels Holm of Odense University found breast cancer cases, split as follows:

|            | MZ | DZ |
|------------|----|----|
| Concordant | 5  | 4  |
| Discordant | 47 | 90 |

Holm concludes: 'A higher risk of breast cancer could be demonstrated in both MZ and DZ co-twins compared to the normal population. The concordance rate was highest in the MZ twin pairs but not significantly different from the DZ twin pairs, so it might be concluded that genes do not determine the development of breast cancer to any great extent.'

Note the use of the word 'might'. Holm goes on to say that his concordance rates were smaller than those obtained in other, earlier studies; perhaps a real genetic influence was masked by the small size of his sample. Further, he also found that, after breast cancer had been developed in one MZ twin, the likelihood of its development in the other twin was six times what would be expected if the two twins had been unrelated (for DZs it was twice as likely). On the other hand, neither twin has an increased risk of cancer at other places in the body.

The Holm study also reinforced earlier work showing that women who marry early and have children are less likely to develop breast cancer. But whether common genetic factors account for all these trends, or whether something to do with early marriage (or maybe early breast feeding of babies) guards against the later development of breast cancer, no one seems to know. Certainly the *number* of childbirths has nothing to do with breast cancer risk.

Finally, the Holm study supported earlier results that women in the same family tend to get breast cancer and to have it in the same breast. But again there was no difference between MZ and DZ twins, so it is hard to see just how the genetic influence is working or how a common environment – being in the same family – could produce this effect.

I started this section by saying that most cancers do not appear to be simply inherited. It can now be seen that the word 'simply' is important. The concordance rates for twins on most cancers are fairly low. But there does, in many cancers, appear to be a 'family effect'. A study in Virginia, by Merton Honeymoon, and presented at Jerusalem, compared the Virginia Twin Register with the Virginia Tumour Register. No increased risk of cancer in young (twin) children was found (up to age fourteen), which seems to rule out the fact of multiple birth as a causative factor. The Danish Twin Register has just gone into cooperation with the Danish Cancer Register (which has a computer file on all malignancies diagnosed in Denmark since 1943). They intend to apply the classic twin study method to the incidence of certain cancers: breast, uterus, stomach, bowel, leukaemia and malignant lymphoma. A study in Connecticut is also looking at cancer in the mothers of twins, based on the theory that they have higher levels of gonadotropins than other mothers.

I personally suspect this is an area where our understanding will increase quite a bit.

## Other illnesses

At first sight, it seems impossible to tell whether the normal infectious diseases of childhood are genetically based, since all or nearly all children at some stage get measles, mumps, chicken pox or whooping cough. But MZ and DZ twins provide once again a natural laboratory. They are the same age and share the same family, so both are exposed to whatever illnesses are going. If genes play no part there should be no difference between MZs and DZs. Luigi Gedda and his colleagues from the Mendel Institute in Rome put this to the test. Their sample consisted of 656 pairs of twins: 199 MZ, 386 DZ and seventy-one of unknown zygosity. Studying the differences between DZs and MZs they came to the conclusion that some people *are* more susceptible to contracting illnesses than others, and that it differs from illness to illness. The heritability of the illnesses they looked at were:

|                 | *Percentage* |
| --------------- | ------------ |
| German measles  | 86.35        |
| Chickenpox      | 30.92        |
| Measles         | 25.44        |
| Mumps           | 5.04         |

To round out the picture, here is the NRC's list of diseases, broken down according to degree of heritability. Remember, though, that even a highly heritable disease may affect only one member of a DZ twin pair, and often is not passed on from an affected parent to his or her children.

### Diseases which are highly heritable

Haemophilia (slow clotting of the blood and excessive bleeding)
Phenylketonuria (a chemical disorder leading to mental retardation)
Albinism (lack of colour in hair, skin or eyes)

Achondroplasia (short stature due to deficient growth of
   long bones)
Huntington's chorea (jerky movements due to specific
   brain damage)
Congenital dislocation of the hip
Strabismus (eyes not lined up correctly)
Some mental deficiency

*Somewhat heritable*

Harelip or cleft palate
Diabetes (sugar not used by body)
Rheumatic fever (damage to heart because of earlier
   infection)
Psoriasis (scaly red patches on the skin)
Pyloric stenosis (narrowing of the outlet of the stomach)
Multiple sclerosis (breakdown of parts of the brain and
   spinal cord)

*Not very heritable*

Congenital heart disease (defect, present at birth)
Headache

## Personality

Personality is the first of several contentious areas which have
been widely studied yet from which a very incomplete picture
seems to have emerged. I am using the word 'personality' in
its widest sense, referring not just to temperament or character but also to such aspects of behaviour as phobias and
crime. Occupational choice and occupational success, which
might also be considered aspects of personality, are treated
later, in the section on intelligence and cognitive functioning.
One other caveat. I can give no more than an introduction to
this subject: according to one estimate, there have been 'doz-
ens of personality scales administered to thousands of twin
pairs in more than thirty studies'. One psychologist who reviewed all these estimated that MZ twins resemble each other
overall 52 per cent of the time and DZs 25 per cent. But at
Jerusalem it was argued that the true figure is half this. A
brief qualification before we look at the studies presented to

the Jerusalem conference: we should bear in mind that, in personality especially, the relationship between genetic influence and environmental influences is complex. That is because there are, in effect, two types of environment. There is the environment *within* the twins' own family – the effect the parents have on the children and the twins have on each other; and there is the environment *outside* the family, plus the fact that families differ. This can make it very difficult to tease out genetic influences from environmental ones. Now let us look at the more interesting studies reported at the Jerusalem conference.

First, let us take the study of 105 same-sex twins from the Louisville Study (same-sex is important in personality studies for obvious reasons). Adam Matheny Jr and Anne Dolan gave the mothers of these twins (who were between seven and ten years old at the time) a personality questionnaire that tested twenty-three different characteristics. These included: whether the twins were daring, impatient, grouchy, artistic, witty, self-disciplined, and so on. Typically it has been shown that identical twins are more similar on these things than DZ twins, but Matheny and Dolan probed more areas of personality. (The possibility arises of course that all the study was reporting was what the mothers wanted to see – and they may have wanted to see identical twins as more similar than DZs. It is difficult to know for sure. But we have already seen that this may be an unfounded fear: the Canadian research mentioned in chapters 4 and 5 supported the idea that parents respond to and do not create similarities in their twins. On the other hand Robert Plouin, from the Institute for Behavior Genetics in Colorado, reported a study at Jerusalem which did not use self-reports (he used videotape instead) and he found *little* or *no* genetic component in personality.)

Anyway, the Matheny and Dolan study showed that identical twins were considerably closer on virtually all of their scores than the DZs. More than that, the *patterns* of scores were more similar for identicals than for DZs. MZs were especially alike in their tendency to be compliant towards rules and regulations, or in their readiness to take risks. The same was true for artistic inclinations, sense of humour, and whether or not they had a wide range of interests.

Gregariousness, too, seemed to be shared more by the identical twins than the fraternals. Being quick-tempered, grouchy and tense also went together.

Obviously, the identical twins were not completely identical in their personalities. But the Louisville study, like many others, supports the idea that quite a strong genetic component underlies personality, presumably acting on our brains in some way, and affecting the basic organization of our behaviour. British studies described at Jerusalem provide further evidence for the genetic influence on personality variables like extraversion – and this was even the case when the British tests were used in Italy. A Texas study also found some personality variables – like Person Orientation – to be linked to *blood group*, a very surprising result and one not yet confirmed.

The Norwegian psychologist, Sven Torgersen, has used a different approach. He has looked deliberately at the *differences* between the personalities of twins. For instance, he looked at 299 twin pairs in Norway where one – but only one – of the twins had been admitted to a psychiatric hospital or clinic for a mild disturbance. He then went back to their childhoods to see how these differed for the two twins.

Torgersen calculated a 'childhood similarity' score based on whether the twins had the same friends, played a lot with each other, were spoken to as a unit, were dressed alike, put in the same class and so on. The most important of his results showed that, as you might by now expect, MZs were more similar than DZs, but the similarity of their childhoods was *completely unrelated* to later differences in personality. So far then, Torgersen's results support the earlier findings. Personality does not appear to be related only to experiences in childhood. Otherwise there should be no differences between MZs and DZs and childhood similarity should have produced some effect on later personality.

However, admissions to psychiatric hospital gave a slightly different picture. For DZ twins their early environment did not have much effect; their concordance on psychiatric admission was the same irrespective of how similar their childhood environment was. For MZ twins, however, the concordance was much lower when the childhood environment was very different. This seems to suggest that, for neurosis at least, there is an interaction between genes and environment.

In another study Torgersen went further in trying to link early childhood experiences with later personalities. Concentrating this time entirely on MZs, fifty in all, he interviewed them about their earlier years (at the time of the study the twins were all between twenty and seventy). He found that the first-born of a pair of twins, or the shorter of the two, were less likely to develop phobias. Dependence in early childhood, passivity and anxiousness lasted into adult life but did not necessarily affect adjustment. On the other hand, the neurotic twin child grew up into the neurotic twin adult. The twin who was dominant as a child grew into the healthier adult. Usually (but not always) this was the first-born. Torgersen's conclusion: 'Modesty in children seems to be the starting point for a general neurotic personality development, characterized by anxiety, insecurity, rigidity, and aggression inhibition.' In a third study Torgersen also found that the submissive twin, the most anxious and insecure, also had a higher blood pressure when examined as an adult. (We have already seen how this may be related to heart disease.)

This appears confusing. Torgersen seems to say at one moment that differences in childhood do *not* affect personality overall but then tells us that yes, the more submissive, more anxious twin grows up neurotic. Perhaps what it boils down to is this: sharing friends, or the same classroom, or being dressed the same does not have an effect anywhere near as lasting as whether you are the dominant or submissive one in a twin pair. Perhaps some twins solve this fairly equitably and both grow up healthy, whereas for others it remains a problem all their lives.

Another study supports this idea. It was found that some temperamental characteristics in children were more genetically based than others and that they varied from age to age. The finding was confirmed both in Britain and Finland, where the heritability of extraversion and neuroticism was found to be significant in the younger age groups. But it was also found that the traits of moodiness and adaptability were the least genetically based – the ones most affected by environment.

It was put forward at Jerusalem that this may be why these are the traits most relevant to behavioural problems in childhood. In other words a poor home or school environment has more effect on a child's moodiness or adaptability than on other aspects of his personality – extraversion for example.

## Fears

In the field of personality research more progress may only be made when the psychologists themselves make up their minds to get away from paper and pencil tests. One interesting project that does this has recently been begun at Oxford by Lindon Eaves, who has come up with some preliminary evidence that the age at which someone first has sexual intercourse may be influenced to some extent by genetics. But by far the most fascinating areas of personality are those of phobias and crime.

Sven Torgersen, the Norwegian psychologist cited in the last section, studied the phobias of ninety-nine same-sex twin pairs, both MZ and DZ. He found that phobias fall into five categories. These are:

| Types of fear | Examples |
| --- | --- |
| Separation | Journeys, shops, crowds, traffic, travelling alone, open spaces, being alone at home, being a passenger in a bus or on a train |
| Animal | Most common were mice, rats, frogs, toads, insects |
| Mutilation | Hospital operations, wounds, injections, the smell of hospitals, pain, doctors |
| Social | Eating with strangers, being watched working, writing or trembling |
| Nature | Sharp objects, tunnels, fire, bridges, mountains, cemeteries, elevators, cliffs, heights, the sea |

Comparing the differences between MZ and DZ twins, Torgersen was able to calculate that all except the first category, separation fears, have a strong genetic component. It was also the case that MZ twins more often feared the same kind of situation than DZ twin partners.

Again he noted that the second-born, often the weaker individual, was heavily represented in some categories, this being particularly true of animal, social and nature fears.

Dr Richard Rose, of the University of Indiana, whose work we have already encountered several times, has also studied fears. He believes they can be divided between those that are culturally acquired (like a fear of God) and those that are a sign of biological adaptation (like a fear of snakes). He gave a 51-item questionnaire to 2000 people – twins and their families – and finds, he says, that the biologically adaptive fears have a genetic component but not the others. His conclusion: '[Our] results are consistent with the suggestion, proposed by Darwin in 1877, that common contemporary fears represent genetic consequences of ancient dangers.'

Still at the 'intriguing idea' stage, Richard Rose also found that MZ twins with two placentas are *no more alike* on certain IQ and personality tests than are DZs, whereas MZs with one placenta are much more similar. Rose speculates as to whether cell division or vascular variation is responsible for this. If his results are confirmed, quite a few of our notions about the differences between MZ and DZ twins may have to change.

Studies with twins also seem to suggest that bedwetting and nailbiting may be genetically based (in one study, four out of six MZ pairs wet their beds). (This same study also suggested that reading disability, constipation and car-sickness may all have some inherited component.) As for sleep-walking, nine out of nineteen MZ pairs were concordant compared with one out of fourteen DZs. On the other hand, MZ children and adults tended to be discordant for tics. Several of these findings strike a chord with Dr Bouchard's results at Minnesota.

Just as intriguing is the result by Dr Fulker and others at the Institute of Psychiatry in London. In a study of 'obsessionality' among 450 twin pairs they found that the most highly inherited aspect of this trait was what they call a 'clean and tidy' factor. It will be recalled that Bridget and Dorothy were both 'obsessively neat' about the house.

## Crime

The position of crime in twin studies is an unusual one. Twenty years ago the view was fairly widely held that genetics played a part in the aetiology of crime, but this approach fell out of fashion. Only in the last few years have new studies been carried out in this area.

On the face of it you might expect criminality to be an easy thing to study. After all, people are either convicted of a criminal act or they are not, they go to prison or they do not. But not everyone gets caught – far from it. Then there are the circumstances of the crime: is a one-time rapist more or less criminal than a man with a string of petty thefts to his credit? Psychologists and criminologists have usually hidden behind the view that a criminal conviction is the 'most objective' criterion.

The main studies of twins and criminality are these:

| Country | Date | MZ | | DZ – same sex | |
|---------|------|------|------|------|------|
| | | No. of pairs | Percentage of concordance | No. of pairs | Percentage of concordance |
| Germany | 1929 | 13 | 76.9 | 17 | 11.8 |
| Holland | 1933 | 4 | 100.0 | 5 | 0.0 |
| USA | 1934 | 37 | 67.6 | 28 | 17.9 |
| Germany | 1936 | 31 | 64.5 | 43 | 53.5 |
| Germany | 1936 | 18 | 61.1 | 19 | 36.8 |
| Finland | 1939 | 4 | 75.0 | 5 | 40.0 |
| Japan | 1961 | 28 | 60.7 | 18 | 11.1 |
| Finland | 1963 | 5 | 60.0 | – | – |
| Denmark | 1968 | 81 | 33.3 | 137 | 10.5 |
| Norway | 1976 | 31 | 25.8 | 54 | 14.9 |

Many of the studies are very old and use very small samples. Although the percentages vary quite a bit, in general there is a higher concordance among MZ twins than DZs. But the last two studies, including the one using the largest sample, found much lower concordance rates than the others.

The Danish and the Norwegian studies should be more representative, as the researchers had access to twin registers, unlike some of the earlier ones, which just took any twins the researchers could find in prisons. The Norwegian study also found very little concordance in the types of crimes committed (theft, fraud, incest, etc.) and no significant difference on this between DZs and MZs.

Among the discordant twins, the Norwegian study found that in most cases the criminal twin belonged to a lower social class, and had more often been physically or mentally ill during his or her life, and more often classified as an alcoholic, than his co-twin. These differences apply to MZs as well as DZs, so they have to be ascribed to environmental or peri-natal factors, not genetic ones. The Norwegians concluded that the predominant factors influencing both criminality and things like alcoholism and poor health were environmental, *later on* in life. (In the only study I have come across which examined juvenile delinquency in twins, no genetic predisposition was found.)

*If* criminality is indeed an aspect of personality (and there is evidence that it is an amalgam of several traits, each inherited to a certain degree), then the figures suggest that to a modest degree (perhaps 25 to 30 per cent) criminality is inherited. But I know of no family studies of criminality among twins, whereas criminologists have identified 'criminal families'. No one suggests that this family syndrome is genetic – the father's teaching is often too evident for that – but it might well affect the results obtained by the twin researchers.

To sum up: personality seems to be moderately inherited. Twins should share a number of traits – something in the order of 20 to 30 per cent, occasionally more. But the relationship between twins, especially along the dominant–submissive dimension, may develop to the point where differences overwhelm and obscure the similarities.

## Schizophrenia

The picture with regard to schizophrenia is as clear as anything in the psychological sciences dares to be. Various studies
have found these concordance rates:

| | | Percentage of concordance | |
| | | --- | --- |
| Country | Date | MZ | DZ |
| --- | --- | --- | --- |
| Germany | 1928 | 67 | 0 |
| USA | 1934 | 67 | 10 |
| Sweden | 1941 | 71 | 17 |
| USA | 1953 | 86 | 15 |
| England | 1953 | 76 | 14 |
| Norway | 1966 | 38 | 14 |
| USA | 1972 | 27 | 5 |

All sorts of reservations can be made about these figures;
they.are, for instance, very variable. And, as with concordance rates for crime, the more recent studies give much lower
figures than earlier ones. By and large, though the concordance rates for DZ twins vary up or down in much the same
way as for MZs, most scientists now accept that MZs are five
times more likely to be concordant for schizophrenia than
DZs. This indicates a considerable genetic component in the
disease, but one which leaves the environment to play an
important role. (Incidentally, MZ twins reared apart show a
very high concordance rate for the illness. Only one proper
study has been done – the Minnesota team hope to do another
– and found the concordance rate was 62.5 per cent.)

The genetic action for schizophrenia may be exerted
through a variety of substances – platelet monoamine oxidase,
platelet gamma amino butyric acid and a host of other can-
didates; but another possibility could be obstetrical difficulties. A study at the Maudsley Hospital, reported at Jerusalem,
found that, in twins whose co-twin had not survived childhood
(34 per cent of the sample, which is very high), the rate of
illness from schizophrenia and other functional psychoses was
twice as high as for twin pairs where both had survived. In

fact, fully 50 per cent of those twins suffering manic-depression had lost a twin.

The most exciting work in this field at the moment, and certainly the most fascinating, is being done by Charles Boklage and his colleagues at the University of North Carolina. Using twin studies, they have found a link between schizophrenia and whether someone is right- or left-handed. Their basic findings are these:

Where one member of an MZ twin pair is schizophrenic, and the other is not, one or both of them is left-handed

MZ twins where both are right-handed are 92 per cent concordant for schizophrenia

Twin pairs with two right-handers have more severe illnesses; those with a left-handed twin are less severely ill

In other words, left-handedness seems to protect at least one twin from schizophrenia and to alleviate the severity of the illness, even when it is present. It is important to say that in those pairs where only one was left-handed it was not always the left-handed twin who was ill; it is left-handed*ness* as such, in twins and not in the individual, that counts.

Twins of both types are more left-handed than the rest of us – but MZs and DZs usually differ in this. Among non-twins schizophrenics are more likely to be left-handed but the difference is by no means as marked as it is in schizophrenic twins. Boklage connects this with the fact that MZ twins may actually take one of three forms, according to whether they share the same chorionic membrane or the same amniotic membrane.

Each foetus is normally surrounded by two membranes: an inner amnion which is filled with fluid for the protection of the baby and is linked to the umbilical cord; and an outer chorion which forms the foetal part of the placenta. MZ twins, it appears, may share the same amniotic sac (in which case they have the same chorion, too), or there may be two amniotic sacs enclosed within one chorion, or thirdly each amniotic sac may have its own chorion. No one is certain as yet but the number of amniotic and chorionic membranes appears to depend upon when separation into two individuals takes place. The earlier this is, the more likely there are to be two amniotic sacs and the more chance the twins have to be slightly different from one another; the later, the more likely they are to have one amnion and one chorion, and the more chance they have of sharing things other than genes, such as the way early division of cells takes place, which may influence laterality (handedness). This fascinating experiment may be the first to suggest that some identical twins are more identical than others. The later the split the more identical they are.

The other category of serious mental illness, the manic-depressive disorders, is very similar in its action to schizophrenia, as revealed through twin studies; concordance rates are around 70 per cent for MZs and 20 per cent for DZs (again the figures vary according to the studies). And even concordance studies for neurosis show it to have a fairly strong genetic component: 60 per cent concordance for MZs (ranging from 25 per cent to 70 per cent according to the study) and 30 per cent for DZs (15 to 43).

## Intelligence

In marked contrast to schizophrenia, scientists cannot seem to agree about intelligence. Some say it is primarily inherited, others that it all depends on upbringing. At times this disagreement has taken extremely unpleasant forms. Psychologist Arthur Jensen of Berkeley, California, has had his offices burned; psychologist Hans Eysenck, from London University, had his hair pulled and spectacles broken during a lecture on race and intelligence in 1973. Some argue that intelligence tests do not measure intelligence at all but some artificial entity created by psychologists. This book is about twins, not the IQ debate: yet we cannot shirk it entirely, since the issue is important and relevant to our main theme.

The fact is, however, that, when we look round at our friends and colleagues, we can usually agree on who are the bright ones among us and who are less gifted. Now this opinion agrees strikingly well with the results of intelligence tests – irrespective of the class or race of person we are talking about. It is also the case that IQ *does* correlate with success in life – certainly far better than any other test or appraisal method yet devised, like the in-depth interview. This fact, that IQ test-scores by and large agree with common-sense notions of intelligence, is and always has been, a major fly in the ointment for those who try to argue that IQ is a synthetic concept. One can advance endless caveats about whether some ethnic groups are penalized by the forms the tests take, or whether the tests neglect certain attributes like 'creativity' – still the fact remains that, within (known) limits, IQ tests measure what most of us instinctively mean when we use the word 'intelligence'.

To what extent is intelligence inherited? Two types of study throw light on this. The first type is illustrated by a survey which pooled the results of several studies. By doing this, material could be used on 122 pairs of MZ twins who were separated at birth and raised apart. As we have seen, the prenatal and immediate postnatal environment even of MZ twins is not absolutely identical but no one seriously thinks these differences significant enough to distort the overall picture. The average IQ of twins in this pooled survey was calculated as 97, which means that they are fairly representative of a normal population. The mean difference in IQ between them was 6.6 points, which works out as a correlation of 0.82. We cannot, however, take this as a true estimate of the genetic component of intelligence without assuming that the twins were brought up in families that were quite random in their characteristics – i.e. not in families of the same class, or living in the same kind of places. Unfortunately, the only study to allow for this was the one carried out by Sir Cyril Burt and now known to have been fraudulent. We must ignore its results.

We do know, however, from Susan Farber's book, which also pooled results of several studies, that most separated twins come from fairly poor backgrounds – which is why they are separated – and that they tend to be raised in lower-middle-class homes or by relatives. In other words, twins who are adopted are not 'assigned' randomly to different types of household: most are brought up in *similar* socio-economic circumstances. Many psychologists would argue that this would make their IQs more similar than might otherwise be the case. Dr Farber also shows that, when the evidence is examined properly, many twins in earlier studies were not separated *completely*: they did see each other from time to time, perhaps every year or so, or on holidays. So she divided the twins in her pooled sample into three groups: totally separated (three sets only), moderately separated, and partially separated. Interestingly, she found that the more separated the twins were, the more unlike were their IQs. She concludes that, at least in part, the similarity in IQ between MZ twins reared apart may be due to the fact that many of them were not truly separated. This is an important caveat, and one that the Minneapolis team are bearing in mind in their own analysis. Dr Farber also found that the results were more similar when the psychologists measuring the twins' IQ knew why they were doing the tests. Bouchard contracts out his testing to professionals who are not told why they are doing it.

So we have to turn to those studies which compare MZs and DZs. The correlations between the IQs of MZs reared together is 0.87, whereas for DZs it is somewhere between 0.50 and 0.55. This gives some notion of the size of the genetic forces shaping intelligence. (And a British study outlined at Jerusalem seemed to support the idea of assortative mating – that is, the tendency for brighter mothers to marry bright fathers. If this is so, the genetic influence on IQ would be underestimated by most studies.) As with the other topics we have considered, there is plenty of evidence for the genetic influences on IQ. For instance, the resemblance of children to their parents suggests that high IQ is a dominant characteristic – that is, the genes which carry it are in some way 'fitter' than ones which do not; if you possess both low IQ and high IQ genes, the high IQ genes win out. This makes sense, of course, in an evolutionary context, for the species needs as many high IQ people as it can to survive.

While this issue continues to exercise the minds and energies of many psychologists, sociologists and biologists, twin researchers for their part seem to regard the issue as very largely settled: at least, to judge by their papers at Jerusalem, they are no longer interested in researching the matter. Two other areas took their interest.

One was that, by holding genetic influence constant, through twin and adoption studies, they could seek to examine the environmental influence on cognitive abilities. One of the studies, from the Institute for Behavior Genetics at Colorado, found, after surveying a number of experiments including the Colorado Adoption Study, that environment accounts for only 10 per cent of the variance in cognitive abilities. It is an important study with somewhat bleak implications. Another study from the same stable found that opposite-sex twins tended to have lower IQs than anyone else – as if something about this configuration hampers the children.

Two other studies, in Canada and Australia, examined the differences between twins and non-twins. The Canadian study confirmed that twins were subject to greater stress early on in their life – lower birth weight, more breech deliveries, more time in hospital – as we saw in chapters 3 and 4. But it also found that twins had lower verbal IQs accompanied by lower test scores in school abilities like reading comprehension and maths. More twins than singletons had failed exams and were undergoing speech therapy. Looking to the future, however, perhaps the most important result of this study is that though the standardized tests showed sizeable differences between twins and singletons, the teachers' ratings in the schools concerned showed *no awareness* of the differences. This could mean that teacher blindness, if indeed this is what is being revealed, may make the problem worse.

One other finding is worth mentioning. These twins had recovered from their (relative) physical weakness by the age of nine. From then on they were indistinguishable from non-twins. However, their psychological disability remained but seems to have stemmed from language deficiencies which emerged much earlier, from as early as two in fact. One wonders whether this is inevitable or whether, just as the physical problems are overcome, so too can be the psychological ones.

The Australian study, at La Trobe University, is a much more elaborate affair. There, David Hay and Pauline O'Brien are investigating 1000 children – twins, their brothers and sisters and cousins – mostly followed from birth. The study is psychological, medical and social. They are also very interested in the difference between first- and second-born twins. So far, they say, their twin–sibling comparisons 'are overwhelming'. There are important differences on all but two tests (this is for twins aged six to fifteen).

Twins do worse than non-twins on most tests of cognitive ability and girls perform better than boys, with male twins doing the worst of all. On a few tests, however, twins outperform their sibs – usually on tests of spatial ability. Second-born twins on average perform more poorly than their first-born co-twin, but the birth order effects were confined to only a few tests. On one, speed of information processing, the second-born did best. Birth order in twins usually does not affect intelligence test scores. But being a twin still does.

This study also found that the mothers were more willing to accept differences between their twins in the areas of personality (and in how 'loving' they were) than in either physical or mental abilities. So IQ differences, between twins and between twins and singletons, could be a continuing source of friction in a family.

Professor Bouchard, at Minnesota, has found that not only are the overall IQ scores of his twins similar, but that the patterns are too. These are early days, but if specific areas of the brain are responsible, roughly, for specific abilities, this twin evidence could help narrow down what these areas might be.

To end this section on an intriguing note: Professor Richard Rose, of Indiana University, is developing a new line of research using the particular genetic consequences of a pair of identical twins marrying another pair of identical twins. The children of the two marriages, who are legally cousins, are biologically as alike as (normal) full brothers or sisters. He is thus comparing these unusual families to test how their IQs vary.

He and his colleagues have found that the 'half-sibs' of maternal twins are significantly more alike in their IQ scores than the 'half-sibs' of paternal twins, which means that mothers appear to have more of an effect on the intelligence of their offspring than do fathers. Since studies about class differences in IQ usually classify children according to their father's job, and neglect the mother, it is not yet clear just what this result means. But it seems as if some modifications of our ideas may soon have to come about.

That completes our brief survey of international twin research. What light does it throw on the main theme of this book – the coincidences between separated twins?

In the last chapter we tried to work out the likelihood that the various coincidences between twins might have occurred by chance. We saw that several things, like phobias, nailbiting or bedwetting, were not as unusual as might be thought and that coincidences of these kinds were evidence of nothing very much, certainly not of any 'uncanny bond' between the twins. We also saw that some similarities *were* more unusual, statistically speaking, and that the odds against them occurring by chance alone were very long indeed. We also noted that many twins shared a long chain of these coincidences and that the probability that a whole sequence might occur by chance was infinitesimally low. We were reluctant to put these links down to astrology or ESP – since there did not seem to be any evidence for such an explanation. So we set out to look at the evidence for a genetic influence on behaviour to see what help that might be. How far has this been proved?

The overall picture is far from clear, but certain (more specific) conclusions can be made with a fair amount of confidence. First, genetics does appear to have a much greater influence on behaviour than is generally thought – from alcoholism to gregariousness, from crime to schizophrenia, from phobias to reading maps (spatial ability).

Next, we have encountered in this chapter research that directly affects some of the bizarre coincidences found at Minnesota. The London study showed 'neatness' to be an inherited characteristic; the Mormon study suggested that even accident-proneness might be inherited, too. And we know that food fads, as well as being common, are also inherited. So these characteristics offer little support for any 'uncanny bond'. Who knows how many other specific traits are inherited in a similar way?

We can also say of twins that five major aspects of their lives are under the influence of genes. We have already seen that their physical health shows a strong genetic influence, and of course they tend to look alike. This helps shape the second aspect, their social life, as indeed do the remaining three factors. Since intelligence is so highly heritable, the earning capacity of twins, their ability to see career openings and to exercise their judgement are also likely to be similar. Their personality similarities add to this: in so far as they like the same things, share similar values, are nervous on the same occasions and so on, their lives are bound to take on similar flavours. Finally, we have touched on the tendency for people to seek out people who are similar as spouses – 'assortative mating'. If twins tend to marry people who are themselves similar in terms of personality, intelligence and other characteristics, it follows that the families of MZ twins are closer than average and this might help account for such coincidences as similar names for their children, same size of families, same number of marriages, similar holiday habits and the rest.

In short, the evidence suggests that what we see at Minneapolis is but an extreme example of the immense influence of genetics on behaviour. However, one of the problems in drawing any general conclusions from chapter 5 was that, in nearly every case, there was just one example of each coincidence so it was impossible to tell, beyond crude statistics, how unusual it was. We know that Oskar and Jack liked to frighten people with sudden sneezes – but we have no knowledge of the incidence of such behaviour in twins, or anybody else. We know Daphne and Barbara fell downstairs, but we do not know how often other people fall downstairs. The evidence in this chapter tells how, on some characteristics, twins compare with the general population. This enables us to say that, in certain areas of life, twins are $x$ or $y$ times closer to each other than the rest of us. That helps us to understand coincidences, and place them in proper perspective.

Take crime as an example. Various studies show that the

concordance rate for crime among MZ twins ranges from 25 per cent to 64 per cent. Let us settle for a figure near the middle as the 'true' concordance figure, say 45 per cent (that is, among twins where one of a pair has a criminal conviction, 45 per cent of the co-twins also have a conviction). Now, in Britain, Home Office figures show that the chances of *any* person being a convicted criminal during his or her lifetime is 31.3 per cent (the Home Office calculates this on the basis of a life-span of eighty years). Therefore, if you stop every person in the street until you get someone who has a conviction, the chances of the *next* person you stop after that also having a conviction are 31.3 per cent. This is the 'concordance' rate for the general population (we will round the figure down to 30 per cent for ease of calculation).

It therefore follows that, on the trait of criminality, twins are closer than the rest of us by the ratio of 45:30 – or, they are one and a half times as close as the rest of us. This calculation may be repeated for a variety of characteristics for which figures are available.

The pioneers of this approach were the Finns; the Finnish Twin Registry was, so far as is known, the first to attempt to measure the relative closeness of twins. The table which follows gives us some idea of this closeness for a number of activities:

| Trait | Percentage concordance in twins | | Percentage freq. in gen. pop. | $\frac{(a)}{(c)}$ | $\frac{(b)}{(c)}$ |
|---|---|---|---|---|---|
| | MZ (a) | DZ (b) | (c) | | |
| Type of work | 55 | 46 | 35 | 1.6 | 1.3 |
| Job status | 82 | 75 | 61 | 1.3 | 1.2 |
| Salary | 65 | 52 | 37 | 1.7 | 1.4 |
| Drinking habits | | | | 0–5 | 1–5 |
| Drug habits | | | | 5–76 | 2–12 |

To take one example from the table and to explain more

fully, 37 per cent of Finns share the same salary band, whatever it is. When one Finn meets another there is a 37 per cent chance their salaries fall in the same range. The chances for two MZ twins are, however, 65 per cent. On salary, therefore, MZ twins are 65/37 = 1.7 times closer than non-twins. The Finns made some attempt to control their figures for sex, age and social class. Below I have attempted to extend their approach in other psychological areas but the available figures do not enable me to allow for these factors:

|  | *Percentage concordance in twins* | | *Percentage freq. in gen. pop.* | $\dfrac{(a)}{(c)}$ | $\dfrac{(b)}{(c)}$ |
|---|---|---|---|---|---|
|  | *MZ* (a) | *DZ* (b) | (c) | | |
| Neurosis | 27–40 | 15–43 | 20 | 1.4–2 | 1–2 |
| Crime | 25–64 | 10–53 | 31 | 1–2 | 1–1.7 |
| Same IQ | 50–82 | 50–55 | 23 | 2–3.6 | 2–2.1 |
| Schizophrenia | 27–80 | 5–17 | 1 | 27–80 | 5–17 |

When you look at the figures in these two tables, quite an interesting picture emerges. Take first the Finnish results. These show that, for everyday things like jobs, where environment comes into play, twins *are* more similar than non-twins but not dramatically so. The 'similarity ratio', as we might call it, is roughly between 1:1 and 2:1. The same is true of neurosis, crime and perhaps IQ, according to the figures of the second table.

Second, when we get to more specific aspects of life, like drinking and drug habits, the extent to which twins are more alike goes up – to between 5:1 and 12:1 (the figure of 76 applies to one drug, in one age group, for women only).

Then we have schizophrenia. The picture here is of quite a different order and probably reflects that, much more than any other characteristic, schizophrenia is very strongly inherited. But it is all part of our picture. We may summarize this by saying that, for most everyday things, twins may be expected to be moderately more alike than the rest of us; but the rarer a behaviour or a trait is the *more likely* twins are to share it. This is because, as the sub-theme of this book has shown, genetics can now be seen to play a much bigger part in our lives than has hitherto been thought. What does this mean for our statistical analysis of the Minneapolis results? Simply this: that on almost *any* aspect of life we should expect at least twice as many coincidences between twins compared with those that the rest of us experience; in the type of job they have, the education they receive, their general interests and so on, coincidences between twins are twice as likely as with the rest of us; with slightly more specific things – their favourite drink maybe, their favourite colour or their clothes – coincidences are, roughly, five to ten times as likely. Finally we can say that the rarer the habits or behaviour patterns you are studying the *more* likely it is that some of these will show up as a coincidence. This may sound paradoxical but it is the picture that emerges from the two tables. Schizophrenia is by far the most unusual phenomenon mentioned here, yet it is one area where the coincidence rate for MZs reaches at times into the 80s. If the coincidences between twins, so far as schizophrenia is concerned, can be eighty times as common as for the rest of us, who knows what other *specific* aspects of our personality or behaviour may be similarly affected? Like flushing the lavatory before using it? Or 'squidging' up one's nose? Or building benches round the tree in one's back garden?

Let us now go back and look again at the calculations about some of the coincidences shared by the twins at Minnesota.

## The Jim twins

To recap:

|                           | Probability |
| ------------------------- | ----------- |
| Both drive Chevrolets     | 0.14        |
| Both bite their fingernails | 0.05      |
| Both heavy smokers        | 0.059       |
| Both heavy drinkers       | 0.16        |

We can now amend these figures to allow for known

concordances. We can assume, for instance, that their choice of car is affected by their salary, which, we now know, is 1.7 times more likely to be in the same range than for strangers picked at random. We may say, therefore, that this coincidence in salary levels is likely to make the chances of both twins buying the same car 1.7 times more likely than if they were two unrelated people picked at random, increasing the probability to $0.14 \times 1.7 = 0.238$. Their drinking habits, as we can see from the table on page 183, are five times as likely to be shared as non-twins'; they are twice as likely to share smoking habits. Finally, for the sake of argument, let us take it that, in their case, nailbiting is an expression of some neurotic tendency: we know that twins are, roughly speaking, twice as likely to share neurotic tendencies as people picked at random – and the revised calculations for the Jims therefore would look as follows:

| Chevrolets | 0.14 | × | 1.7 | = | 0.238 |
| Nails | 0.05 | × | 2 | = | 0.1 |
| Drink | 0.059 | × | 5 | = | 0.29 |
| Smoking | 0.16 | × | 2 | = | 0.32 |

Multiply these together and you get: 0.002. That is to say, these four coincidences would occur by chance in these twins once in 500 times – rare, but only ten times as rare as the likelihood that the US President will be assassinated in the coming year.

Before we end this chapter, we can make the same amendments for the other twins examined in the last chapter.

*Barbara Herbert and Daphne Goodship*

| Both fell downstairs aged 15 | 0.0006 |
| Both have a fear of heights | 0.38 |
| Both drink vodka | 0.11 |
| Both drink black coffee | 0.15 |
| Both wear similar clothes | 0.0008 |

There are no figures that enable us to alter the probabilities of the first and last characteristic with accuracy. But, given that their fear of heights is a common neurosis, we can amend the other probabilities as follows on the same basis as we did for the Jims:

| Falls   |        |     | = | 0.0006 |
|---------|--------|-----|---|--------|
| Fear    | 0.38   | × 2 | = | 0.76   |
| Vodka   | 0.11   | × 5 | = | 0.55   |
| Coffee  | 0.15   | × 5 | = | 0.75   |
| Clothes |        |     | = | 0.0008 |

which works out at 0.000,000,15 or 1 in 6,666,666.

*Bridget Harrison and Dorothy Lowe*

| Both wore rings | 0.025 |
|---|---|
| Both wore bracelets | 0.03 |
| Both studied piano and | |
| gave it up at the same time | 0.003 |
| Both were very neat | 0.07 |

Let us try something different this time. MZ twins are clearly very similar in appearance; concordance for certain physical characteristics like eye colour, height and so on is almost total. These are acknowledged to be far more genetically determined even than schizophrenia. This could mean that twins are far more likely to wear similar jewellery than any other two people. Let us assume then that, as with schizophrenia, twins are eighty times as likely to share jewellery-wearing habits as the rest of us. Also, let us take into account the fact that, although we have no precise figure for the likelihood that twins will share similar personalities, we do know they are twice as likely to share the same neuroses. If we can assume that much the same is true for personality this could throw some light on their piano-playing habits, as could their similar IQ level. We also know from the London study that twins are, roughly, two and a half times more alike on neatness than the rest of us. Our revised calculation for these twins would therefore be:

| Rings    | 0.025 | × 80 |       | > 1 (i.e. absolute certainty) |
|----------|-------|------|-------|-------------------------------|
| Bracelets| 0.03  | × 80 |       | > 1 (i.e. absolute certainty) |
| Piano    | 0.003 | × 2  | × 3.6 | = 0.02                        |
| Neatness | 0.07  | × 2.5|       | = 0.18                        |

We have two absolute certainties here so we only have to use the last two figures in our calculation: 0.02 × 0.18 = 0.0036 or 1 in 277, slightly less likely than your chances of ever being in a mental hospital.

These calculations are 'forced', of course, because the figures are so imprecise. Further, there are many other alleged coincidences in the lives of these twins. Not all of the coincidences, even several of them together, are as rare as we may think.

The coincidences shown by some twins are far more remarkable than those shown by others. Even with Barbara Herbert and Daphne Goodship the enormous figures attached to their coincidences are not necessarily evidence for ESP. The chances of getting a hand at bridge made up of the entire suit of spades is 1 in 635,013,559, 599, and when it happens it makes the headlines. But of course the chances of getting *any other* bridge hand are exactly the same – and some sort of hand obviously comes up every time bridge is played. So odds of this magnitude come up all the time – this is the hard point to grasp. The odds on Daphne and Barbara sharing the characteristics we have been looking at are about a million times as common as a particular bridge hand coming up at any one time. Yet whole suits of spades do come up – and no one suggests ESP or some other paranormal force is at work.

Am I being really fair to the parapsychological case? After all, neither the Jims, nor Barbara and Daphne, nor Bridget and Dorothy shared only four sets of characteristics – far from it. Each shared *many* more which, on the face of it, seem to have an infinitesimally low probability of occurrence due to chance alone.

Two points, however, should be made. By no means all the Minnesota twins showed quite the bewildering array of coincidences as Oskar and Jack did, or Barbara and Daphne, or the Jims. Keith and Jake, Ethel and Helen, Dan and Mike, Margaret and Terry, each shared far fewer coincidences. If parapsychology is particularly strong among twins, why is it stronger in some than others? Why do some twins show more coincidences than others? The combined probability of the (relatively few) coincidences seen in these other twins, if it could be worked out, might not be all that low; these twins might be very poor evidence for anything very much, especially parapsychology.

More important, which of the coincidences at Minnesota are really independent of each other? We do not know, for instance, whether people who drink vodka also tend to drink black coffee, whether people who wear blue shirts with epaulettes also tend to flush the lavatory before using it, or whether collecting cuddly toys is in any way related to piano-playing. Such links are not far-fetched: men who work in poorly paid jobs and like hamburger fast-food restaurants, *do* buy the cheaper make of car and holiday on the cheaper beaches; neurotic women of the same age who are afraid of the sea *are* also afraid of heights and small, enclosed spaces; women of the same age, build and IQ level have roughly the same spending power and tastes, and therefore share similar wardrobes – especially in an age of standardized, mass-market clothes (Daphne and Barbara both bought theirs at Marks and Spencer). In other words, we cannot tell (because the research has not been done) to what extent several behavioural idiosyncrasies may reflect far *fewer* underlying genetic similarities.

And even the twins who shared fifteen to twenty unusual characteristics may be showing no more than the behavioural effects of a much smaller number of genes.

This is a plausible line of reasoning, more plausible than ESP or telepathy and not simply because there is absolutely no evidence for these phenomena in connection with MZ twins. We should remember three things:

First, the twins research is taking place against a background of increasing advances in the science of behavioural genetics, which are showing that many more aspects of our psychology have at least some genetic component, compared with what has hitherto been thought.

Second, the most recent research shows that several of the specific coincidences observed at Minneapolis are indeed genetically determined. Among these are neatness, phobias, accident-proneness, musical ability and interests (such as carpentry), to mention only a few.

Third, two kinds of coincidence have been found at Minneapolis, and so far we have dealt with only one. The first type of coincidence relates to shared characteristics – two people both go on holiday to the same place, wear jewellery in the same way, have the same drinking habits or the same fears. The second type has not yet been dealt with but takes us back to the beginning of our analysis of coincidences in general: it concerns not just whether a feature is shared but also introduces the factor of time. Not only did Barbara and Daphne both fall downstairs, they did so when they were both aged 15; both Jims put weight on at the same time, for no obvious reason, and then both took it off again, equally inexplicably. Most notably, Dorothy and Bridget kept diaries for the same year, and Margaret and Terry got married on the same day, at the same time.

We need to consider this phenomenon to complete our investigation of the twin bond and in doing so we shall encounter what is in my view one of the most intriguing subjects in the behavioural sciences – and something that should convince everybody that what we are seeing at Minneapolis is a manifestation of a genetic and not a paranormal influence. This is the subject of 'chronogenetics', the way our genes control not only the way we are at birth but also the way we develop throughout our lives – our puberty, the spurts and lags in our intellectual growth, the ageing process, even (to a certain extent) when we will die. Understanding the last two of these phenomena must be of the greatest interest to us all.

# 7 The New Science of Chronogenetics

## Clark Gable dies on the dot

In November 1960 the film star Clark Gable was admitted to hospital after a heart attack. While he was there George Thommen, an American scientist, said on the Long John Radio Program, 'If I were Mr Gable's doctor, I would have everything that I possibly could have in his room for the greatest emergency on the 16th of this month.' Mr Gable died on the 16 November.

Mr Thommen was not an astrologer or a clairvoyant, at least not in the traditional sense. He was an advocate of 'biorhythmics', a branch of physiology and psychology which claims that our bodies go through various cycles, regular peaks and troughs, which make certain days more suited to particular activities and others especially risky.

The concept of cycles within nature is an old one. Linnaeus, in the eighteenth century, even constructed a 'flower clock' which began at 5 a.m., when the common sowthistle opens, and continued every hour until 9 p.m. when the night-flowering catchfly opens. Many aspects of our body have a cycle, very often lasting about a day: our blood pressure follows such a rhythm, with peaks in the afternoon, and our kidneys do too, being less active at night. Men's testosterone peaks at seven in the morning which is why, contrary to the way most people behave, this is the best time of the day for a man to have sex, whether to enjoy himself or to conceive children. Illnesses, too, show rhythms. Strokes and asthma tend to occur at night, epileptics are more likely to have fits between 6 and 7 a.m., and less likely between 5 and 9 p.m. Malaria attacks every second or third day. Psychoses often show rhythms over a cycle lasting a few days, a month or even a year. Peritonitis, oedema, purpura, mouth ulcers and migraines are all known to occur at fixed intervals.

So the idea that physical and psychological states should have some rhythm to them is by no means new.

Biorhythmics extends this idea, however, into a more general principle: that there are three separate cycles within our bodies which control our *physical, emotional* and *intellectual* states, each on a different time-scale. Their 'discovery' can perhaps be put down to the old German charlatan we came across in chapter 5, Dr Wilhelm Fliess. Fliess, you may remember, believed in two cycles which governed men's lives, one of 23 and one of 28 days. His evidence was very thin and he has been totally discredited now; but other doctors, most notably the Japanese, the Americans and the Swiss, claim to have found evidence that not only do we have a physical cycle of 23 days and an emotional one of 28 days, but that there is also a third, intellectual, cycle of 33 days. These run through the months as shown in the figure opposite and a great deal of work has now gone into the effects which the cycles may have on people's lives.

Followers of biorhythmics believe that a person's cycles start at birth and continue through life; they also believe that, as one passes from a positive phase of a cycle into a negative one, or vice versa, a 'critical' period is encountered, variously physical, emotional or intellectual according to the cycle involved. Because the cycles are fixed, we can work out in advance the days on which people's criticals fall, especially the days when more than one critical occurs at the same time, when we are in special danger. Factories in Japan and Switzerland claim to have saved thousands of francs and yen, as well as some lives, by giving workers days off when they have multiple 'criticals' and therefore are specially accident-prone.

George Thommen had worked out from Clark Gable's birthday (1 February 1901) that he would have a physical critical on 16 November 1960. That is why he made his 'prediction'. There are many other cases where biorhythms have seemed to work.

On the other hand, the medical director of Swissair, which has often been said to be one of those companies which uses

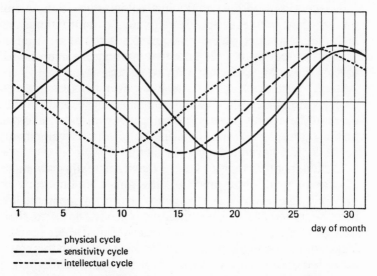

——————— physical cycle
— — — — sensitivity cycle
-------- intellectual cycle

One biorhythmic month. Note that only ten days are *not* within forty-eight hours of a critical (after *Number Power*, by Keith Ellis)

biorhythms, denies this. The company did study the system, he says, but found nothing in it. As the doctor says, if you examine a biorhythm chart there are in fact only ten days in the month which are not within two days of some 'critical' or another.

I mention biorhythmics for two reasons. First, if there is something in it, and our cycles do start at birth, then twins should have identical cycles and so reach 'criticals' and even 'double' or 'triple' criticals at the same time. Biorhythms could increase concordance and this might explain such co-incidences as Daphne and Barbara falling downstairs at the same time (though we do not know if it was on the same day, just the same year). On the other hand, the same argument would apply as much to DZ as to MZ twins and we have already seen, in our discussion of astrology, that DZ twins do not show anything like the coincidences that MZs show. So the twin evidence is against biorhythms, much as it is against astrology. (Starting the cycles from the time of con-ception rather than from birth is no help, as the same objec-tions apply.)

A second reason for mentioning biological cycles is to emphasize that, whatever the shortcomings of biorhythmics itself, the idea underlying chronogenetics is not at all strange. There is a great deal of evidence that our bodies are governed, at least on a day-to-day basis, by some sort of internal clock. A longer-term clock operates in so far as our bodies undergo changes at puberty, then later at the menopause, and at other times in the ageing process. Chronogenetics tries to take this further, looking at specific aspects of our development, psychological development especially.

Earlier chapters showed how MZ twins actually grow more alike as they get older, and DZs less alike. No one thinks it very odd that even MZ twins who have been separated for forty years should still be similar physically, down to their weight, teeth and crooked little fingers. But if their psychological development is parallel in a similar way then this could be an important reason why some of the coincidences in the lives of separated twins occur.

The crucial study of spurts and lags in the growth of twins was carried out by Ronald Wilson, Professor of Paediatrics at the Child Development Unit at the University of Louisville School of Medicine in Kentucky. Wilson looked at the development of 374 pairs of twins and their siblings; they were tested initially when they were only three months old, then every three months for the first year, every six months in the second and third years, then annually until they were six. The children were given a whole range of intelligence and development tests.

He found, as many studies have, that the abilities of MZ twins were much closer than for DZs. But far more interesting results were obtained when he looked at individual pairs of twins. As the graphs in the figure opposite show, the MZ twins went up and down in their abilities together. Some pairs did better as time went by, some did worse – but in all cases *both* twins in a pair went in the same direction. Even more impressive, the profiles of the pairs were similar: if one twin lagged in a particular test, so did his or her co-twin; when there was a spurt, it affected both. The same was not true, at least to anywhere near the same extent, among DZ twins, as the graphs also show.

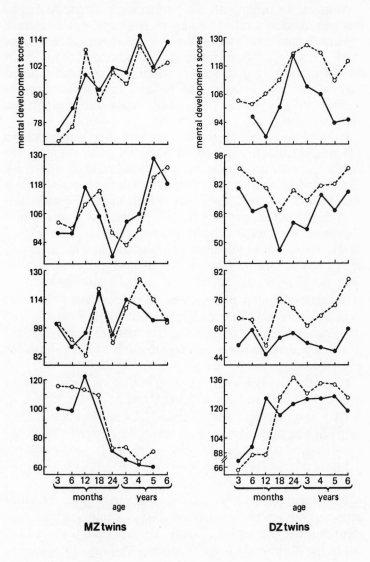

Spurts and lags in the mental development of twins. Not only
are MZ curves closer than DZ curves, but spurts and lags
are very similar too (after 'Synchronies in mental development:
an epi-genetic perspective', *Science*, 1 December 1978

Although to begin with the performance of one MZ twin was only modestly related to his or her co-twin, this picture stabilized (for everybody) around the twenty-four-month mark. To Wilson this seems to suggest that parts of the brain do not mature until this time, when the baby has learned basic things like walking and general coordination and is turning to the more demanding tasks of talking and understanding.

Wilson also looked at the experiences of twin pairs where one was premature (judged by very low birth weight). He found a big difference in their performances in the early months, which by the twenty-four-month mark had virtually disappeared. So there does seem to be an in-built resilience in young children, perhaps helped by the fact that the first two years is set aside for physical rather than psychological development, which together ensure that a child handicapped in this way at birth has the chance to catch up.

In the same study, Wilson found that spurts and lags in physical development also showed high concordance but these were *independent* of the changes in mental development. This therefore seems to underline the fact that the spurts and lags in development are not a general growth factor but are under more specific genetic control.

In another study by Hugh Lytton in Canada a similar point is made. Lytton and his colleague Denise Watts looked at thirty-seven pairs of twins (fifteen MZ and twenty-two DZ) between the ages of two and nine. They found that some behaviour, like speech, remained more stable in MZs during this time than in DZs but that other things, like independence and compliance to parents, did not.

The suggestion is clear: there is some sort of genetic control over certain specific aspects of our psychological development. Further supporting evidence comes from studies carried out at the other end of our life-span. We have already come across one such study, establishing that twins, especially MZs, are far more likely than either brothers or sisters to die at roughly the same time as their co-twin – that is, they have approximately the same life-span. The latest study of this kind is especially interesting because it concerns a very detailed set of records – from the Mormons in Utah in the United States.

## The Mormon death study

The Mormon Church has unusually complete records and the Mormon Genealogy Data Base was recently computerized by the University of Utah. From its 1.2 million records Dorit Carmelli and Sheree Anderson were able to collect 2,267 sets of twins born between 1800 and 1899, in other words all Mormon twins born in the nineteenth century – a pretty impressive achievement. They followed the lives of these Mormons and recorded when the various twins died, according to whether it was during infancy (the first year of life), as youngsters (under 44), in middle age (to 65) and above that. Not only did twins who died at an old age tend to die near one another, but the same picture was obtained with the other age groups, even the very young. In other words there was strong evidence of a genetic influence at whatever age the twins died. Environment played a large part but by no means gave the whole picture. Genetics too were highly significant.

## Ageing is not dying

Then, finally, there is the intriguing work of Professor Philip Burch at the University of Leeds in Great Britain. Professor Burch is an authority on the process of ageing and is known among the academic community for his view that ageing and death are not one and the same process. For example, baldness and going grey are both examples of ageing but neither is known to be lethal or life-shortening. His view (and this is a brief summary of course) is that ageing is due to the body turning against itself and destroying itself after its resistance has broken down. If he is right, we may eventually see the process of ageing stopped.

However, it is not this but another aspect of Professor Burch's research that concerns us now: his research into heart disease. He has noticed that, after the first appearance of heart disease, there is a 'latent period' when nothing much appears to happen; after that time, the disease recurs, usually fatally. Two points need to be made about this. In the first place, the latent period is different for particular genetic groups. In white men, for instance, it is on average twelve years, whereas in women it is twice that, i.e. twenty-four years. For black men it is different again.

This latent period, therefore, appears to be under genetic control – and so here is another area where genes govern our behaviour over time.

Professor Burch thinks that this pattern may actually exist in other conditions but has not yet been noticed. The latent period may be a time of especial risk, when what happens to a person – what his or her environment presents – is as important as genes. We know, after all, that if someone takes it easy after a first heart attack, they can survive much longer than if they continue to pursue the same pace that brought about the attack. If Professor Burch's view is correct, then the concordance rates for many diseases in twins are actually *higher* than reported because, after the first episode, they react differently, with different outcomes, and so the true concordance is underestimated. Professor Burch believes that several studies of twins, that intended to throw light on the role of genetic predisposition in various illnesses, may have seriously underestimated and misrepresented the role of genetics in the past. (I should add that some researchers argue the opposite and believe that chronogenetics may exaggerate heritability.)

The research in this area is still too much in its infancy for us to be able usefully to amend our calculations about twins and their degree of relative similarity. But it does serve to remind us of two things. Most importantly, that all our calculations so far may actually have *underestimated* the degree of similarity between the Minnesota twins and therefore overestimated the odds of the various coincidences. Secondly, the research makes it doubly clear that coincidences in time need to be looked at with much more care before we can conclude that there is something truly uncanny (in the sense of being inexplicable) about them.

One further point might also be made. The fact that spurts and lags in development may be under genetic control could also mean that there are times when we are especially vulnerable to the environment. If this were so then, by and large, twins would tend to be more similar than any other two people; but, in those cases where one twin's vulnerability was exploited and the other's was not, the differences between them could become more important than the similarities. By the same token, small changes in the environment, if they took place at such vulnerable times, could produce considerable effects, whereas at less vulnerable periods even large environmental changes would be relatively less significant.

# Conclusion
## *Is the bond what it seems?*

Many people have essentially irrational attitudes to statistics and to what they regard as unusual occurrences. I hope the figures in chapter 5 have convinced the sceptical reader that rare events really *do* happen without the intervention of any supernormal phenomenon.

To date, there is no evidence whatsoever to support the idea that any form of parapsychological phenomena are involved in the twin bond. Admittedly, very few studies have been done but the results have all been negative – there is not the slightest scintilla of a suggestion that twins have some way of communicating with each other that brings on coincidences. Or, at least, if there is, the twins themselves know nothing about it. Several of the twins in the Minnesota study were not even aware they were twins beforehand and, they say, had no sense of being 'incomplete' or lacking. Indeed, one or two were actually annoyed when they found they were twins; Daphne Goodship, at least, although she knew she was a twin for ten years before her sister sought her out, never had any desire to go looking for her 'other half'.

I have given little space to astrology in this book, although a number of reputable psychologists (like France's Michel Gauqelin and Britain's Hans Eysenck) have recently given more respectability to the subject through studies which appear to show that the season of birth may have something, however vague, to do with personality and, through that, with someone's choice of career. The reason I have spent so few pages on astrology is simply that the whole body of twin evidence goes against it. The very fact that *all twins*, MZ or DZ, are conceived at the same time, yet the two types grow up in very different ways, seems to me the most convincing evidence that astrology, if it has any influence at all on personality, has only a very slight one. (Maybe twins are not the most powerful evidence: triplets may be better. We quoted earlier studies of triplets which consisted of MZ twins and a third singleton, in effect a DZ twin to each of the other two. Here we find that the MZs grow alike but that the odd twin out is very different. Astrology, it seems to me, cannot account for this.)

Nor should we forget that the twins in the Minnesota study, even those with some of the most surprising coincidences, also showed many differences. But there is at work here a mechanism which appears to maximize the similarities and coincidences between twins at the expense of the differences. For the Minnesota study to have lasting significance, some acceptable *definition* of a coincidence will have to be found, so it can be checked whether the similarities really do outweigh the differences. Scientifically, the only way of measuring coincidences of the sort that gave rise to this book would be to construct a list *beforehand* and then compare the number of coincidences between each pair of twins with those between pairs of non-twins. It would have to be a pretty long list with some fairly specific questions (like: Do you read magazines back to front? Do you flush the lavatory before using it? Do you giggle a lot? Do you push up your nose and call it 'squidging'?). Clearly it would leave out a lot of other coincidences that twins new to the study might have in common. But that brings us back to the definition of coincidences and the extent to which we pay attention to them when they occur.

Everyone has been fascinated, for instance, by the fact that some of the twins wear the same clothes, of one sort or another. Yet everyone must have been in the same situation himself – I have. At least four times in my life I have worn the same shoes as someone else, or the same tie. Twice I have been with women who had on the same dress as another woman at the same function. Yet no one thinks this odd or anything more than a coincidence.

Perhaps what is going on is the 'halo effect', a manifestation of Dr Philip Zimbardo's 'small world phenomenon'. He has found that if you take two ordinary (and totally unrelated) people in, say, the United States, they are on average only five stages of friendship from one another. That is to say, any person knows someone who knows someone who knows someone who knows someone who knows someone who knows the other person. Diagrammatically, it can be represented like this:

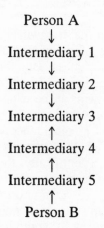

Person A

↓

Intermediary 1

↓

Intermediary 2

↓

Intermediary 3

↑

Intermediary 4

↑

Intermediary 5

↑

Person B

You can observe this in action anywhere in Europe any summer's day. Two Americans arrive (separately) at a monument, museum or pavement café. Tired from their sightseeing, they notice from the other person's dress and manner that he or she is American. They fall to talking. . . .

'Where are you from?'

'New York.' (Or California, Kansas or Seattle.)

Then begins the search for friends or experiences in common.

'I was drafted in Seattle once. Used to drink in O'Lunney's on 23rd Street. Know it?'

If the other person knows it, pretty soon they find acquaintances in common. If not the conversation moves on to other areas of potential common experience: college, more often than not, or jobs they have had, rock concerts they have been to and so on. Before long, however, our two holiday-makers unearth someone they have in common, and go away from the encounter muttering, 'Small world, isn't it?'

The truth is, of course, that, if each person has only fifty acquaintances (and who has not if you include family, school and college friends, neighbours and workmates?), and if those fifty acquaintances each have a further fifty and so on, it only takes five intermediaries to clock up 312 *million* people ($50^5$) or almost half as much again as the population of the United States. So, it is *virtually certain* that you will have a mutual acquaintanceship, through perhaps four or five intermediaries, with *anyone* you meet on holiday. When on top of that you actually look for things in common, by saying where you were at college, or where you were in the army, or where you had your honeymoon, you go beyond your fifty best-known acquaintances and take in all sorts of other people. Either way you are bound to say, 'Small world, isn't it!'

Something analogous may be going on with the coincidences in the lives of separated twins. We are all made up of a myriad different characteristics. As Keith Ellis has pointed out, the man who tells his girlfriend, 'You're one in a billion' is not necessarily flattering her. It could be a shrewd appreciation of the odds. For instance, she could have:

|  | Probability |
|---|---|
| a Grecian nose | 0.01 |
| platinum-blonde hair | 0.01 |
| each eye a different colour | 0.0001 |
| a good knowledge of maths | 0.0001 |

If you multiply these together (and we are neglecting her height, complexion, the shape of her lips, her bust, her beautiful neck) we end up with a combined probability of 0.000,000,000,001, which is one in a million million.

With a sufficiently long list of characteristics it may simply be that we *all* share a few with any other person, so it is not surprising at all to find that the separated twins share characteristics. Maybe just because they are twins and we have grown to believe that they have so much in common, we automatically notice the coincidences in their lives and neglect those in our own. This is a question we cannot answer either way at the moment: the necessary research has not yet been done.

We looked at the statistics for coincident birthdays earlier.

Let us look at that particular coincidence. The average life expectancy is 69.5 years. So the chances that any person you bump into anywhere in the world was born on the same day as you is 1 in 25,384, since that is the number of days in 69.5 years. Those may look like long odds but remember there are 4,000,000,000 people alive on earth to choose from. So there are, roughly,

$$\frac{4,000,000,000}{25,000} = 160,000$$

people alive today around the globe who were born on the same day as you (rather more if you are young, rather fewer if you are old). Seen from that angle is it any surprise that the Wallaces in their *Book of Lists* cite no less than 117 pairs of *famous* people born on the same day. These include:

| | |
|---|---|
| 9 January 1941 | Joan Baez and Susannah York |
| 30 January 1937 | Vanessa Redgrave and Boris Spassky |
| 12 February 1809 | Charles Darwin and Abraham Lincoln |
| 3 April 1923 | Marlon Brando and Doris Day |
| 23 June 1894 | Alfred Kinsey and the Duke of Windsor |
| 24 August 1872 | Aubrey Beardsley and Max Beerbohm |
| 1 December 1935 | Woody Allen and Lou Rawls |

Consider also the children of Ralph and Carolyn Cummins of Clintwood, Virginia, USA. On 20 February 1952, Mrs Cummins gave birth to their first child, Catherine. Curiously, their second child, Carol, was also born on 20 February a year later. Imagine their feelings when their third child, Charles, entered the world in 1956 – on 20 February. But this was not the end. Claudia Cummins was born on 20 February 1961 and lastly Cecilia in 1966 – on 20 February. All five children, four girls and a boy, were born on the same date. The odds against this are 1 in 17,797,577,730 – roughly four times the world population.

*Rare events do happen.* If the Minnesota findings are confirmed to show a rate of coincidence far higher than chance allows, its cause must be explained by some genetic mechanism. And a genetic explanation is by no means an unexciting one. If even our little habits or specific preferences – for white seats around the tree, or for blue shirts, for Alistair Maclean novels or chocolate – are in some way inherited, then the way we think about ourselves is going to alter radically.

None of this is known for certain yet, but in recent years the science of behaviour genetics has been making much headway. To begin with, its methods were somewhat inaccessible to the layman. Experiments were usually conducted on mice, the aim being to see if there was such a thing as a 'neurotic' mouse. This is judged in mice, apparently, by how frequently they defecate and it was found you can breed a strain of mouse that is inherently nervous: it fouls its cage much more than well-bred stable mice. New ideas and new statistical techniques have enabled behaviour geneticists to examine more real problems and to attempt an understanding of human behaviour. The trend, on all fronts, is towards the view that genetics exerts a much stronger influence over behaviour than anyone ever thought. In one recent paper the British authors actually state: 'Studies of normal and abnormal personality traits [suggest] social environment to be unimportant . . . individual differences in personality may have an entirely constitutional base.' Intelligence was seen by these authors as not being quite so bound by our constitution, but heredity nevertheless still outweighed the environmental influence substantially.

The Minneapolis research is part of this trend. Now that scientists are aware of the significance of genes for behaviour, they are looking for this influence – and finding it.

Susan Farber, in her book *Identical Twins Reared Apart*, drew attention to the many shortcomings of research before the Minnesota study. Specifically she found that IQ tests used were out-of-date or not standardized on major groups, that much of the evidence was anecdotal, and that the trauma of separation for the twin children may have contaminated the results (for example, it could have made their dreams rather similar if the separation was reflected in them). She also noticed that twins who had been reunited before being studied were *less* similar than those who had not; a process of 'twinning' had undoubtedly taken place. She found too that the spouses of twins were *not* similar, as might be expected. All these shortcomings were in addition to those listed here in the Introduction, and they amount to a major new caveat in evaluating studies of twins reared apart. Nevertheless, Dr Farber's conclusions – apart perhaps from those relating to IQ and allergies, which she found to be fairly discordant – are not so different from mine, despite her criticisms of the scientific standards of the studies she looked at. For instance, she also concluded that the *patterns* of twins' lives appear to be similar, and she lists a long string of anecdotes about attributes, habits and gestures of MZ twins reared apart which are difficult to measure but appear to be strikingly similar (two of the twins in her pooled sample were called John, like the Jims in the Minnesota study; two others, Earl and John, took up the violin, like Dorothy and Bridget playing the piano; two women preferred *Messiah* to any other music, rather in the way that several sets of twins at Minnesota liked the same authors).

It is too early to say where this growth in the influence of behaviour genetics will lead. Many people – scientists, doctors and teachers – feel threatened when any results point in this direction. They fear they will be used to justify a particular set of (usually right-wing) policies. Yet the geneticists and psychologists engaged in behaviour genetics are usually only intellectually interested in the problems they study, and rarely have a political axe to grind.

The last twenty years have seen a paradox in the social sciences in the West. In many countries – America, Britain, Scandinavia especially – social scientists have, for the first time, found themselves in a position to implement policies based on their studies and theories. Yet many of the problems which those policies were designed to eradicate have in fact grown. Without a doubt policies on crime, suicide, violence, race relations and much else have failed. Policies designed to eradicate inequality in education have failed spectacularly. Have these failures occurred because social scientists have misunderstood human nature? Have they assumed that change can be brought about by environmental manipulation when people are less malleable than they would like to believe? On the other hand, policies relating to mental health – involving the use of drugs – have been very successful. Is this because the genetic base of the problem has been acknowledged, and led doctors to look for medical treatment rather than for help from the 'talking', manipulative therapies?

It could be argued that the Minneapolis research is but the most specific and colourful example to date to show that our genes govern a large part of our lives, including areas we never imagined could be so affected. But this is *not* a 'fascist' or even reactionary conclusion. It *does* mean that we may have to accept certain inherent limitations within ourselves – not just in our abilities, but in our interests, our likes and dislikes, and so on. It may mean accepting – in a broad, crude sense – the unpalatable fact that class differences owe something to heredity, as well as to 'breeding'. But it means much else. Behaviour genetics seems to teach us that human babies have a couple of years at the start of life which are a kind of 'cushion', during which the main aim is physical development: intellectual impairment caused in that time may be reversible. Ronald Wilson's work with 'half-sibs' suggests that there is something in the quality of mothering, as opposed to fathering, which helps young children. Feminists may not like the look of this, but it is clearly a potentially important phenomenon, if further research sustains Wilson's early results.

In other words, we are not faced here with a plain 'either/ or' contest between genetics and environment for the chief role in determining human behaviour. The situation is much more complex, and we are only now beginning to understand the interactions. This understanding surely provides new opportunities for knowing ourselves better and may perhaps prevent us from entering some of the blind alleys into which past social experiments have led. Though the evidence is still tentative, and few facts can be sure in the realm of behaviour, this is the trend now to be derived from most modern research.